# PRACTICAL
# SURVEYING
# FOR
# TECHNICIANS

# PRACTICAL SURVEYING FOR TECHNICIANS

**Robert P. Landon, C.S.T.**
Member A.C.S.M., N.S.P.S., L.S.A.W.

Delmar Publishers Inc.
I(T)P™

## NOTICE TO THE READER

Delmar staff:
Sponsoring Editor: Mark W. Huth
Project Editor: Danya M. Plotsky
Senior Production Supervisor: Larry Main
Senior Art and Design Supervisor: Susan C. Mathews

For information, address

Delmar Publishers Inc.
3 Columbia Circle, Box 15015,
Albany, NY 12203-5015

Printed in the United States of America
Published simultaneously in Canada
by Nelson Canada,
a division of The Thomson Corporation

1  2  3  4  5  6  7  8  9  10  XXX  00  99  98  97  96  95  94

Library of Congress Cataloging-in-Publication Data

Landon, Robert P.
    Practical surveying for technicians / Robert P. Landon.
        p.    cm.
    Includes bibliographical references and index.
    ISBN 0-8273-3941-0
    1. Surveying.  I. Title.
TA545.L25  1994
526.9 – dc20                                                  92-40713
                                                                  CIP

# Contents

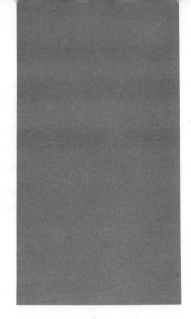

# Preface

This text is written for the technician. It is written in trade language, for easier understanding, and is intended for use as a reference book by the survey technician who is preparing to take promotional examinations.

As the practice of surveying requires constant upgrading of the skills learned, this book has been written with a minimum of questions to allow the instructor to add recent innovations without having to explain why the questions in the book are out of date.

The text leads the student through a survey problem, from the initial planning and field reconnaissance to the final design and map. This will allow the technician who is primarily interested in either field or office work to become familiar with what is required for a complete survey project from start to finish. Whether or not the student intends to continue and become a licensed surveyor, the text will show that the technician is capable of contributing to the overall projects engaged in by a surveying firm.

The trade terms used in the text are the ones that I have learned in my many years as a party chief in the field. If local terms are different, please substitute the proper one for your area.

This book could not have been written without the help of my students, who have taught me more about the fine points of surveying through their questions than I would have thought possible. I also thank the members of the Land Surveyors' Association of Washington who generously shared their wisdom while supplying many of the old instruments and books that helped make up the chapter on survey history. I am obliged, also, to the eagle-eyed reviewers, whose valuable suggestions and corrections helped me avoid several slips.

I acknowledge John Thalacker for loaning the antique surveying equipment that has been used in some of the figures in the text.

I extend a special thank you to my family for letting me trespass on their time to help with the research: to my son Don Landon for his work in the darkroom, and to my daughters and grandson who drew some of the illustrations in this book. I owe more than I can say to my wife, Darlene, who not only put up with me during the writing of this book but did the word processing and organization work. Without the help and encouragement of all these wonderful people, I would never have finished.

# Introduction

There are many excellent textbooks written for use in four-year degree programs and there are also elementary books for the property owners who want to know about where their property line is located. However, the students in my surveying technology class needed a book for the entry level survey technician that was written by a survey technician. After reviewing many texts without finding what I was looking for, the Delmar representative suggested I write my own textbook. I did and this is it!

My thirty years of field experience covers areas throughout the West Coast of the United States, Alaska, Mexico, and the Marshall Islands. It includes public land surveys, construction, hydrographic, tunnel, lot, subdivision, pipeline, and whatever else needed to be done. When work was slow, I filled in as a heavy equipment operator and grade checker. This gave me an opportunity to learn about how others used the stakes we had set for grading. In this text, I will pass on to you as much of that experience as I can.

The lessons will take a subdivision survey from start to finish. The skills needed are taught in the first chapters; the rest of the book covers how to put those skills to use. It takes only a short time to learn *how* to take a rod shot, but it takes many weeks of practice to learn *where* to take that shot. There is no one to check your work in the field, so I have not given the answers to the questions in the textbook. Making the proper mathematical checks of your work will prepare you for the conditions you will find in the field and office. The ability to THINK is the most important tool you will have at your disposal. Being able to solve problems independently is essential. You may be working far from help and there will be no one to ask. When you have successfully completed all the lessons in this book, you will at least be familiar with most of the situations you will encounter when you go to work as a survey technician.

GOOD LUCK!

# A Proud Tradition: A History of Surveying

## Objectives

After completing this chapter, the student should be able to:

1. Relate the history of surveying in the United States.
2. Discuss, critically, how the history of surveying in the Old World influenced surveying in this country.
3. Name several of the well-known surveyors who performed surveys of historical significance in the United States and identify the surveys for which they are known.
4. State why it is necessary to understand how the original surveys were performed in order to do retracement surveys.

Welcome! By choosing to study the trade of surveying you are entering an honorable profession that traces its roots back to the Neolithic Period. Ancient peoples found the need to make maps of trade routes and settlements. As the idea of ownership of property developed, it became necessary to mark permanent boundaries. In 2000 B.C., Babylonians laid down the first foundations of property law in the Law of Hammurabi.

The ancient Egyptians used a system of angles and distances to replace the land marks (monuments) that were wiped out by the annual flood of the Nile River. In the tomb of Menna, at Thebes,

1

**Figure 1-1** Harpendonaptae or rope stretchers like those in the painting from the tomb of Menna. *Courtesy of Darlene Landon*

there is a wall painting showing two chainmen (Harpendonaptae or rope-stretchers) surveying a field of corn (see Figure 1-1). One theory on the pyramids supports the notion that they were used as primitive **theodolites** to show directions for the resurvey of the flooded fields. The accuracy of the measurements for the Great Pyramid of Cheops is still a marvel to modern surveyors.

We credit the Egyptians with the invention of geometry to lay out the boundaries of the land and to make **maps** that would record the ownership. They also made maps of the surrounding country; a map of the Nubian gold mines still exists today. The Etruscans developed a system of **rectangular surveys** that are similar to those that we use today in the United States Public Land System.

## THE ROMANS

The Romans used surveyors, called agrimensores, to lay out the system of roads that allowed the Roman legions to move rapidly from one part of the Empire to another. Along these roads, cities grew where the armies had built forts to garrison the conquered lands. With cities, the need for water led to the construction of aqueducts, many of which still stand today. Surveys were used to lay out the city streets and mark the land **corners**.

Many of today's surveying instruments have evolved from those of the Romans. They used the groma (see Figure 1-2) to lay off right angles. A 10-ft **rod** with bronze ends was used to measure distances; a Roman foot was 13.2 inches. An instrument similar to the plane table was used for mapping.

In A.D. 79, a Roman general, Gnaeus Julius Agricolo, brought surveying to Britannia (present-day England) when he built a system of forts and roads to allow him to move his legions into Scotland. The cities of Colchester, Lincoln, York, Glouchester, and St. Albans stand on sites originally surveyed by the Romans.

Cities were laid out in rectangular blocks, intersected at right angles by streets of various widths: a system used in ancient Babylon by the Etruscans. It was passed on to the Romans who, in turn, spread its use through Gaul (Europe) and Britannia. After the Romans lost power in England, the Anglo-Saxon system of land ownership — based on the whims of the King — returned.

In 1066, the Norman Conquest brought a system of feudal land laws that are the basis of the land titles we have today. The concepts of fee simple and the possibility of a life estate were in use. Fee simple endured as long as the tenant or any of the tenant's heirs (whether the descendants or not) survived. Life estate ended with the death of the tenant. In 1285, the third estate of fee tail was created, which was good as long as the tenant or any of his descendants lived. These are all based on the English theory that a tenant could not own the land, only an estate on that land.

**THE NORMAN CONQUEST**

The Statute of Frauds (1677) mandated that transactions of land be in writing. With written deeds came the necessity of a written description of the land conveyed, and the surveyor became a respected local figure.

**THE STATUTE OF FRAUDS**

The real property laws of the United States came, quite naturally, from the English laws. The Colonial charters provided that the lands were "free and common socage." This was a type of tenure whereby the tenant held a free title, subject to restrictions. The Colonists were so against the idea of feudalism that attempts by landholders of the crown to collect rents led to incidents such as Bacon's rebellion in Virginia in 1676.

A charter, granted to Massachusetts Bay in 1691, included the Plymouth Colony and the Provinces of Maine and Nova Scotia. The line between Massachusetts and Rhode Island is part of the southerly line of the grant. A dispute over the location of that line went on for over 200 years. The location of the disputed line was argued before the Supreme Court by Daniel Webster and Rufus Choate as counsels for Massachusetts.

**THE COLONIAL CHARTERS**

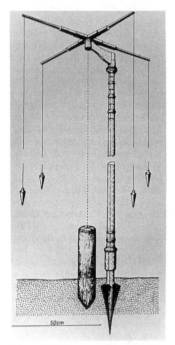

**Figure 1-2** Groma. *Courtesy of Carl Zeiss, Oberkochen, Germany*

In 1642, Nathaniel Woodward and Solomon Saffrey set a stake on the Plain of Wrentham that marked a point three miles south of the Charles River. In 1710, commissioners from Massachusetts and Rhode Island agreed that

> That stake set up by Nathaniel Woodward and Solomon Saffrey, skillful, approved artists, in the year of our Lord one thousand six hundred and forty-two, and since that often renewed, in the latitude of forty-one degrees and fifty-five minutes, being three English miles distant southward from the southern most part of the river called Charles River, agreeable to the letters-patent for the Massachusetts Province, be accounted and allowed on both sides the commencement of the line between Massachusetts and the Colony of Rhode Island.

The Woodward-Saffrey stake is also mentioned in the Connecticut boundary as the "Burnt Swamp Corner."

In 1703, the commissioners of Rhode Island and Connecticut agreed on "A straight line from the mouth of the Ashawaga River to the southwest corner of the Warwick purchase, and thence a straight north line to Massachusetts." Dexter and Hopkins surveyed this line, which is still known as the "Dexter-Hopkins Line."

## THE MASON-DIXON LINE

In 1760, two famous English mathematicians came from England to settle the dispute over the line between Maryland and Delaware. Charles Mason and Jeremiah Dixon verified the line already run and also ran the line between Maryland and Pennsylvania. This line is probably one of the most famous survey lines ever run. Who hasn't heard of "The Mason-Dixon Line," which figured so prominently in United States history? Their work was so well done that 200 years later the northeast corner of Maryland was found to be only 2 sec. off from its intended position.

Mason and Dixon began work in 1763 and, after having run 244 miles of line, ran into a property dispute with the Native Americans, halting work in 1767. A commissioner from Virginia, named Andrew Ellicott, continued the Mason-Dixon line westward in 1784. Ellicott, as will be demonstrated shortly, went on to become quite a prominent figure in the history of surveying.

In 1772, Valentine and Collins ran a line from the eastern shore of Lake Champlain eastward to the Connecticut River as part of the northern boundary of the Province of New York (now Vermont). Part of this line west of Halls Stream to the deepest part of Lake Champlain was to become part of the boundary between the United States and Canada.

**Figure 1-3** George Washington, surveyor. *Courtesy of the Surveyor's Historical Society*

The line was continued westward from Lake Champlain by Collins and Sauthier in 1773. The following year, Samuel Holland and David Rittenhouse (another prominent figure in surveying history) set a stone on a small island in the Delaware River between New York and Pennsylvania. Andrew Ellicott used this stone in his survey of the boundary line of Pennsylvania. The desire to own their own land combined with the outrage of a tax on tea prompted the colonists to call on a surveyor from Culpepper County, Virginia, to lead the Colonial Army in 1776 (see Figure 1-3). George Washington went on to become the first President of the United States.

**A SURVEYOR FROM CULPEPPER COUNTY**

In 1790, Ellicott ran a **traverse** from the west end of Lake Ontario along the shore to the Niagara River, up and across the Niagara, and along the southwest shore of Lake Erie to establish the starting point for running the west boundary of New York.

In 1799, Andrew Ellicott, as Commissioner for the United States, was directed to survey the line between Georgia and Florida from the Mississippi River to the Atlantic Ocean. A boundary

**THE UNITED STATES**

marker that Ellicott established north of Mobile, Alabama, is still in existence. It is the initial point for the St. Stephens base and **meridian** and is also a National Historical Civil Engineering Landmark.

In 1810, the state of Georgia employed Ellicott to survey the 35th degree line of latitude to be the boundary between North Carolina and Georgia. Many other surveyors were also performing surveys during this period but, because of his exceptional skill and the large area covered, Andrew Ellicott is probably one of the most well known of all the early surveyors.

Benjamin Banneker assisted Andrew Ellicott in the survey of the Federal Territory, now known as the District of Columbia. Banneker was a black man skilled in science, mathematics, and astronomy. Perhaps his greatest contribution was the calculation for the ephemerides for the Benjamin Banneker Alamanacs he published from 1792 through 1797.

Ohio was the first state to be created from the Northwest Territory beyond the Ohio River. In 1802, Congress passed an act "to provide for the due execution of the laws of the United States within the State of Ohio." In this way, the great expansion of the United States to the west was begun.

**THE LOUISIANA PURCHASE**

In 1803, the United States purchased the Province of Louisiana from Napoleon for $23,213,567.73 including interest, roughly four cents per square mile. Thomas Jefferson, who was President of the United States at the time, wanted to explore the 830,000 square miles of newly acquired land. On January 18, 1803, he asked Congress for $2,500.00 to pay for an expedition to map the new land and search for a "northwest passage." To lead this expedition, Jefferson chose William Clark, General George Rogers Clark's younger brother, and Captain Meriwether Lewis.

**LEWIS AND CLARK**

In the spring of 1803, Lewis was trained by Andrew Ellicott to make observations that were "to be taken with great pains and accuracy, to be entered distinctly and intelligibly for others as well as yourself." Lewis and Clark left St. Louis on May 14, 1804, with a group of twenty-six soldiers, interpreters, Clark's servant, York, and Clark's Newfoundland dog. They traveled up the Missouri River and across the Bitterroot Mountains and down the Columbia River, reaching the Pacific Ocean on December 3, 1805 (see Figure 1-4).

The Lewis and Clark expedition gave Jefferson the public endorsement he needed to launch his long-dreamed-of exploration of the Louisiana Purchase. Jefferson wrote, "The work we are now

**Figure 1-4** Lewis and Clark map. *Courtesy of the Oregon Historical Society*

doing is, I trust, done for posterity, in such a way that we need not repeat it. . . . We shall delineate with correctness the great arteries of the country: those who come after us will extend the ramifications as they become acquainted with them, and fill up the canvas we begin."

**THE WEST**

Thomas Freeman, the next surveyor-explorer of the West, set out on April 19, 1806, from Fort Adams (near the junction of the Red and Mississippi rivers). Freeman, accompanied by twenty-four men, planned to explore and map the Red River. The Spanish had issued a proclamation prohibiting exploration of the Red, and a previous expedition, led by William Dunbar, had been turned back after 3 months by the threat of Spanish intervention. Freeman hoped to succeed given the large number of people in his group. After exploring only 600 miles, however, the Freeman group was intercepted by a large Spanish force that ordered them to turn back.

Another expedition, led by Lieutenant Zebulon Pike, which had departed from St. Louis in August of 1805 to find the source of the Mississippi River, also had bad luck. Soon after his return, Pike was ordered to lead another expedition, this time to locate the headwaters of the Arkansas and Red rivers in the Spanish-claimed southwest. Pike and his men were the first white Americans to cross from the Kansas to the mouth of the Arkansas River. Pike considered the country he crossed through to be barren. His description of the terrain caused the area to be called "The Great American Desert," and it was believed to be a barrier to further westward migration for many years to come.

After succeeding in crossing the "desert," Pike discovered the peak that was named for him (although he never reached the top because he underestimated its height). On leaving Pike's Peak, the expedition made a winter crossing of the Sangre De Cristo Mountains into what is now New Mexico. Pike and his men were soon picked up by a large troop of Spanish soldiers who escorted them to Santa Fe, where his papers and maps were confiscated. He was then forced to return to the United States through Texas.

When Jefferson left office in 1809, there occurred a period of indifference regarding further exploration of the western territories. In 1812, the Louisiana Territory was renamed the Missouri Territory and was known by that name even after the State of Missouri was created in 1821. After the War of 1812, another wave of westward exploration was begun by Stephen Harriman Long.

Between 1816 and 1823, Long led five expeditions into the Missouri Territory, covering more than 26,000 miles. Long shared Pike's opinion that the West was a great desert and, as a result, further exploration by the government ceased.

Demand for furs re-opened the West to exploration by a group of adventurers that became known as "mountain men." Such names as Joseph Walker, the first white man to see Yosemite Valley; Christopher "Kit" Carson, who acted as guide for Fremont; John Colter, whose description of "Colter's Hell" (now known as Yellowstone National Park), caused disbelief at the time; Jim Bridger, who built a fort in what is now Jackson Hole; William Sublette; Jedediah Smith; and many other brave explorers who "saw the bear." The stories told around the campfires and carried back to St. Louis, along with the furs, dispelled the myth of the "Great American Desert."

In 1834, that part of Missouri Territory lying north of the State of Missouri and east of the Missouri and White rivers was taken by the Territory of Michigan. Kansas and Nebraska Territories absorbed the remainder in 1854. Wisconsin Territory was formed in 1836 from that part of Michigan Territory that lay west of the State of Michigan. All that part of the original Louisiana Purchase and the Red River basin south of the 49th parallel became the Territory of Iowa in 1838.

**OREGON TERRITORY**

Charles Wilkes was only a lieutenant when the squadron of six ships left Hampton Roads, Virginia, in August of 1838 to begin the four-year voyage of exploration. Wilkes's map of the "Oregon Territory," issued in 1844, opened the area to settlement by American settlers and the eventual claiming of the territory by the United States.

In 1842, Thomas H. Benton, a powerful United States Senator and expansionist, secured an appointment for his son-in-law, John C. Fremont. Fremont was to lead an expedition to survey the Platte and Arkansas rivers upstream to the Sweetwater. He chose Kit Carson as his guide. As one of the mountain men who had knowledge of the area, Carson was to become famous as a western hero in many penny novels sold to Easterners who were eager to learn about the West.

Fremont's first expedition accomplished very little in the way of new discoveries; however, it did pave the way for his second expedition in 1843. Fremont was ordered to "join on to your positions of 1842 on the Colorado of the Gulf of California. Thence continuing northwestardly in to the Flathead country, and join on to Lieutenant Wilkes's survey."

After completing his assigned survey, Fremont decided to turn south through California, Nevada, and Arizona. He made a winter crossing of the Sierra Nevada Mountains to Sutter's Fort in California. He then turned east through the Mojave Desert to return to Independence in July of 1844.

The third and best known expedition began in June of 1845. Fremont returned to Sutter's Fort and became active in organizing the American settlers in a revolt against the Spanish. He was appointed military commandant and led the California battalion in the war against Mexico that led to independence for California. During the next thirteen years, the Army's Corps of Topographical Engineers completed over twenty major surveys of the West. With the onset of the Civil War, this group of West Point graduates was absorbed into the Army Corps of Engineers.

As a result of the Mexican War settlement, it become necessary to survey 1800 miles of the new border between the United States and Mexico. This survey was made between 1849 and 1855 under the command of Lieutenant William H. Emory. During this same time, Lieutenants George H. Derby and Joseph C. Ives were exploring and mapping the Grand Canyon. In April of 1858, Lieutenant Ives and his party became the first white men to reach the bottom of the Grand Canyon near Diamond Creek.

Meanwhile, Captain Howard Stansbury and Lieutenant John W. Gunnison were mapping the great basin around Salt Lake. Guided by Jim Bridger, Stansbury and Gunnison blazed a trail across Cheyenne Pass through the Rocky Mountains that was later put to good use by the Overland Stage, the Pony Express, and the Union Pacific Railroad.

## THE BUILDING OF A RAILROAD

In 1844, Asa Whitney proposed the building of a railroad from Lake Michigan to the mouth of the Columbia River. Senator Thomas H. Benton told the Senate, "We must have surveys, examinations, and explorations made. . . ." And who should be in charge of these surveys? Why, his son-in-law, John C. Fremont, of course.

In October of 1848, Fremont set out to explore the 38th parallel. Guided by Old Bill Williams, Fremont foolishly attempted a crossing of the San Juan Mountains in December. Ten of his men died from exposure and starvation. The twenty who survived by eating the mules reached Taos in February.

In 1853, Congress passed the Pacific Railroad Survey Bill. Three survey parties were to survey the "most practical" routes. Isaac I. Stevens, the governor of Washington Territory, led the party surveying the northern route. The second party, led by Captain John Gunnison, was to survey the 38th parallel, Fremont's route. The third, led by Lieutenant Amiel W. Whipple, was to explore the 35th parallel. Meanwhile, Lieutenants John G. Parke, Henry L. Abbot, and Robert S. Williamson searched for passes through the Sierra Nevada Mountains of California.

About this time, the Native Americans began to notice that wherever surveyors went, more white men soon followed. In 1849, the Pit River Indians ambushed a survey party in the Sierra Nevadas. The leader, Captain William H. Warner, and his guide were killed by arrows.

On October 26, 1853, while camped near Sevier Lake, Captain Gunnison's party was attacked by a war party of Utes led by a warrior named Moshoquop. Only four of the troopers escaped; all the rest of the party were killed. Although Gunnison's notes were recovered, the disaster closed this route to serious consideration.

The southernmost route along the 32nd parallel was favored by Secretary of War, Jefferson Davis. In 1854, he sent parties led by Lieutenant John Parke and Captain John B. Pope to survey the route, simultaneously, from the east and the west. It was no sur-

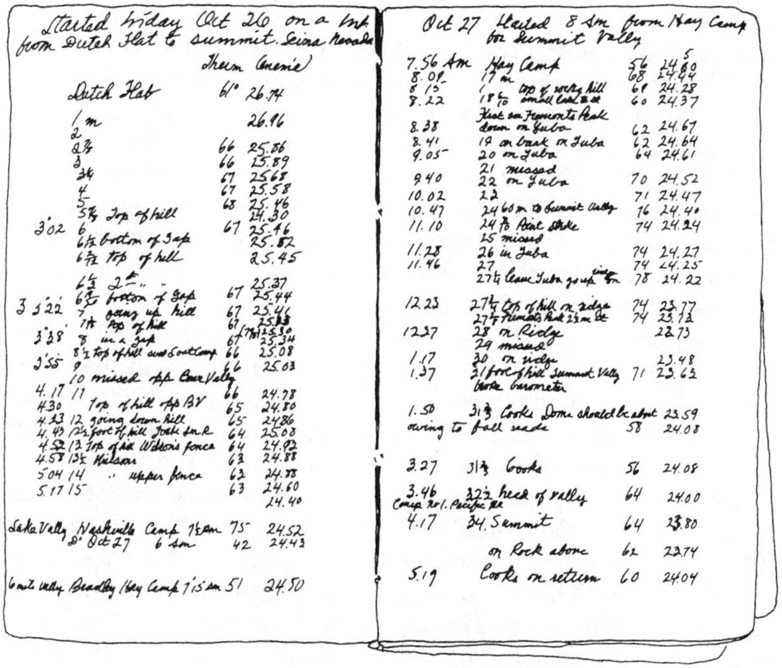

**Figure 1-5** Theodore Juda's survey notes of route through Donner Pass. *Courtesy of Dale Landon Thomas*

prise, therefore, that Davis chose the southern route for the railroad. A major problem with this route was that it crossed through Mexican territory. To solve the problem, Davis authorized James Gadsen to purchase the disputed territory for $10,000,000. Before Davis's plan could be carried out, however, the Civil War broke out, and the Corps of Engineers had other, more deadly tasks to perform. After the war ended, at Appomatox Courthouse on April 9, 1865, the nation again turned its eyes westward.

Theodore Judah was a dedicated railroader and civil engineer and surveyor. It was through his surveys (see Figure 1-5) that the route through the Sierras via Donner Pass was completed. (Judah used a route that crosses just south of Donner Pass. This route was discovered by Daniel Strong, a pharmacist and amateur surveyor from Dutch Flat, California.) In 1869, the Union Pacific and Central Pacific railroads joined at Promentory Point in Utah. They were built along the route pioneered by Fremont, Gunnison, and Judah.

In 1861, Dakota Territory was organized from parts of Minnesota and Nebraska Territories. In 1863, Idaho Territory was made from parts of Washington, Dakota, and Nebraska Territories. Montana Territory was created from part of Idaho in 1864. An unsuccessful effort was made in 1858 to create the "State of Jefferson" out of the present State of Colorado and parts of Nebraska, Wyoming, and Utah. In 1861, however, the Territory of Colorado was created by act of Congress, being composed of parts of Utah, New Mexico, Kansas, and Nebraska Territories. The Territory of Montana was created from Idaho in 1864, and Wyoming Territory was taken from Dakota Territory in 1868.

Oregon Territory was organized in 1848 as "all that part of the Territory of the United States which lies west of the summit of the Rocky Mountains, north of the Forty-second degree of north latitude. . . ." In 1853, Washington Territory was formed from Oregon Territory north of the Columbia River.

Nevada Territory was created in 1861 from part of Utah. Deseret Territory was originally settled by the Mormons in 1847. Congress, however, did not recognize their delegate or their petition, and, as a result, Utah Territory was established in 1850. It was bounded by California on the west, Oregon on the north, the Rocky Mountains on the east, and the 37th parallel on the south.

The Republic of Texas declared its independence from Mexico in 1835. In 1850, the United States paid $10,000,000 for land west of the 103rd meridian, providing for the formation of the Territory of New Mexico. This land, along with part of the land gained in the treaty of Guadalupe-Hidalgo, formed New Mexico Territory. The Gadsen Purchase of 1854 added additional area to the territory and in 1863, the Territory of Arizona was formed from the western part of New Mexico Territory.

**THE HOMESTEAD ACT OF 1862**

The Homestead Act of 1862 opened the land for settlement by anyone over twenty-one years of age or the head of a household to file a claim on not more than 160 acres of surveyed public land. This act created a great land rush to the western territories and the need for more information about the land to be settled. To this end, Ferdinand V. Hayden was placed in charge of the Geological Survey of Nebraska in 1867, under the auspices of the General Land Office. The title was changed, in 1879, to the United States Geological Survey of the Territories under the Department of the Interior.

**THE GREAT SURVEYS**

In 1870, Hayden's party surveyed the area north along the North Platte to the Sweetwater through South Pass and then southwest to Fort Bridger (see Figure 1-6). After working south into the Uinta Mountains, they turned north up the Green River into Wyoming Territory then east through Bridger's Pass to Cheyenne. "For every mile on the map we covered two or three on the ground; up mountain sides, down stream beds, across country; to gather rock specimens, to survey and map, and to paint and photograph," reported expedition photographer, William H. Jackson.

**Figure 1-6** Ferdinand Hayden's survey party. *Courtesy of the National Archives, Washington DC*

The first official survey of Yellowstone began in June of 1871. Hayden and his party added to the information gathered by the Washburn-Langford-Doane party of 1870. He and Langford together managed to convince Congress to establish the first national park, and on March 1, 1872, President Grant signed the bill creating Yellowstone National Park. Hayden continued to survey the west he loved, from Wyoming to Arizona, until he retired in 1878.

Clarence King began his career with the California Geological Survey as a volunteer geologist. Working for Josiah Whitney, he surveyed and explored the High Sierras for four years. King was than placed in charge of surveying a 100-mile-wide path along the 40th parallel from the crest of the Sierra Nevada Mountains to the slopes of the Rocky Mountains. It took King and his party ten years to complete the survey.

In 1879, King was named Director of the U.S. Geological Survey. In the years between 1867 and 1870, the southern part of the

**Figure 1-7** Lt. George M. Wheeler and Army surveyors. *Courtesy of the National Archives, Washington DC*

**Figure 1-8** John Wesley Powell with unidentified Native American. *Courtesy of the U.S. Geological Survey, Denver, Colorado*

Great Basin, south of the King survey, was surveyed and mapped by Lieutenant George M. Wheeler (see Figure 1-7).

As chief of the Geological Survey west of the 100th meridian, Wheeler mapped much of Nevada, Death Valley, and the Colorado River up to Diamond Creek. In May of 1869, John Wesley Powell started from the Green River, planning to explore and map the Grand Canyon of the Colorado (see Figure 1-8).

With nine men in four boats they explored 1500 miles in ninety-nine days. When one of the boats was lost in the rapids, four of his men, discouraged by the danger and hardships of the journey, left the expedition to return overland. These men, who had left only two days before the end of the trip, were ambushed by Native Americans. Three of them were killed.

On his *geographical and topographical survey* of the Colorado River, Powell had his boat, the *Emma Dean*, fitted with an armchair. From this perch, he was better able to observe the river (see Figure 1-9). When the water was calm he read *The Lady of the Lake* aloud to his men.

Powell's map of the river, made by his topographers H. C. De Motte and F. M. Bishop, is now on file in the Library of Congress. Powell's mapping of Colorado, Utah, and Wyoming led to his appointment to serve as the Director of the U.S. Geological Survey from 1881 to 1894.

As the land was being mapped by the U.S. Geological Survey, it was also being surveyed into **townships** and sections by the Deputy Surveyors of the General Land Office, under the direction of the Surveyor General. Edward Tiffen issued specific instruc-

tions on Land Office surveys in July, 1815. They required that surveys were to be laid out using a 66-ft. **Gunter's chain** (Figure 1-10) and a Rittenhouse **compass** (similar to the one in Figure 1-11). The Rittenhouse compass was used until after 1836, when the **Burt's improved solar compass** (Figure 1-12) came into general use.

---

In 1836, the General Land Office was reorganized and, in 1851, it published *The Oregon Manual.* The Act of September 27, 1850, created donation claims of 320 acres to a single man or 640 acres for a man and wife. The Donation Land Claims were designed to encourage settlement of Oregon Territory.

In May of 1851, John B. Preston established the initial point for the Willamette meridian in the hills west of Portland. This

**THE OREGON MANUAL**

**Figure 1-10** Gunter's chain, Burt's solar compass, timber scribe

**Figure 1-11** Rittenhouse style compass

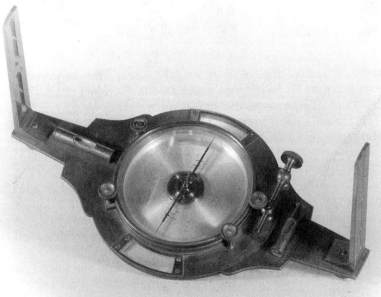

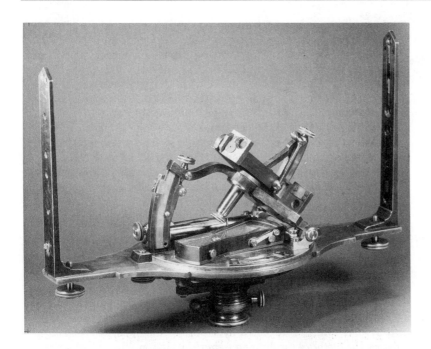

**Figure 1-12** Burt's solar compass.
*Courtesy of Dennis DeMeyer*

point controls all the public land surveys for what is now Oregon and Washington states. Deputy surveyors James E. Freeman and William Ives then ran the line south and north from the initial point. These surveys were made with a solar compass.

On Thursday, July 17, 1851, deputy surveyor Leander Ransom drilled a hole on top of Mount Diablo, California, to establish the Mount Diablo meridian and began the survey of California into sectionalized land. These surveys were made under *The Oregon Manual,* as were the surveys made by Lorenzo Gibson in Little Rock, Arkansas, and George B. Sargent at Dubuque, Iowa.

In March of 1855, John W. Garretson was contracted to survey the New Mexico **principal meridian** using *The Oregon Manual.* Build "a durable monument on the Missouri River," said John Calhoun when he contracted with Charles Manners. That July, a cast-iron post was established 52.55 chains west of the river. The initial point for the 6th principal meridian was established 60 miles west of the Missouri River. These surveys were also made under *The Oregon Manual.*

---

The first *Manual of Instructions for the Subdivision of the Public Lands* was published in 1855. The Manual read in part:

Standard Parallels are established at stated intervals to provide for or counteract the error that would result from the convergence

**MANUAL OF
INSTRUCTIONS**

of meridians, and also to arrest error arising from inaccuracies in measurements on meridian lines, which however, must be studiously avoided. On the north of the principal base line it is proposed to have these standards run at distances of every four townships, or twenty-four miles, and on the south of the principal base, at distances of every five townships, or thirty miles. [This was changed in later manuals to four townships (uniformly)].

Where uniformity of the variations of the needle is not found, the public surveys must be made with an instrument operating independently of the magnetic needle. Burt's improved solar compass, or other instrument of equal utility, must of necessity be used in such cases. The township lines, and the subdivision lines will usually be measured by a two pole chain of thirty-three feet in length, consisting of fifty links, and each link being seven inches and ninety-two hundredths of an inch long. On uniform and level ground, however, the four pole chain may be used. You will use eleven tally pins made of steel, not exceeding fourteen inches in length, weighty enough towards the point to make them drop perpendicularly. In measuring with a two pole chain, every five chains are called 'a tally' because, at that distance, the last of the ten tally pins with which the forward chainman set out will have been struck. He then cries, 'tally,' a cry which is repeated by the other chainman, and each registers the distance by slipping a thimble, button, ring of leather, or something of the kind, from a belt worn for that purpose or by some other convenient method. (*Manual of Instructions for the Subdivision of the Public Lands* 1855)

These instructions were necessary because many of the chainmen were uneducated and could not read or write; however, they could count to ten. The calls to natural features were memorized and given to the recorder at each tally. The chainmen and axemen were sworn by the deputy surveyor to faithfully perform their duties and they tried their best, given the terrain and attacks by Native Americans. In 1838, a party of twenty-five surveyors in Texas was attacked by Native Americans and eighteen surveyors were killed.

### 1890 Chain-Carriers' Oath

We, _____ and _____, do solemnly swear that we will well and faithfully execute the duties of chain-carriers; that we will level the chain upon even and uneven ground and plumb the tally pins, either by sticking or dropping the same; that we will report the true distance to all notable objects, and the true length of all lines that we assist in measuring, to the best of our skill and ability, and in accordance with instructions given us, in the survey of the _____ _____.

In 1848, Professor William M. Gillespie, a civil engineer, wrote that the roads of America were inferior to those of any other civilized country. At that time, the only roads were wagon roads such as the Santa Fe and Oregon trails. With the railroads being built across the United States not yet in full service, the demand for roads became urgent when gold was discovered in 1849. If someone wanted to get to the gold fields, he either joined a wagon train or went by sea around the tip of South America. Some 42,000 "49ers" went cross-country in 1849 alone.

As more wagons crossed the country, small towns sprang up along the way. In turn, these towns became cities, so the demand grew for surveys of towns and lots for sale.

**THE GENERAL LAND OFFICE**

Land was sold to prospective pioneers through the General Land Office (GLO), sight unseen. A lot could be chosen off a township **plat** and purchased before leaving on the long overland trek. Because of the location of springs and other usable water located on the field notes by the GLO surveyors, a section of land could be purchased with at least a chance of finding water and starting a farm or ranch.

The first roads were just tracks along fence lines. The fences were built along the boundries of the farms that were sold as sections of land. This led to eventual destruction of the section corners set by the government surveyors. As the land was sold and resold, a need arose for local land surveyors to mark the lot corners of the subdivided property.

As the farms and ranches became more prosperous, the farmers raised more than they could use and had surplus crops to ship to market. The need in the industrialized east for the cheaper western wheat and beef led to the expansion of the railroad system.

Before a railroad could be built, the route had to be determined by a preliminary survey. An early railroad survey party usually consisted of thirteen to sixteen men (see Figure 1-13). They traveled by horseback and wagon and lived in tents. The locating engineer and the instrumentmen usually slept in the office tent with the survey gear. The chainmen, rodmen, and axemen slept in separate tents, and the cook slept in the cook tent. The wranglers slept close to the stock to prevent theft of the horses. Personal equipment usually consisted of a folding cot, bedroll, and suitcase.

The crew sometimes surveyed as much as ten or twelve miles a day in good country. Camp had to be moved often to prevent losing time in travel. The crew worked ten to eleven hours a day, six days a week, but Sunday was always a day of rest. If the camp was close to the railhead, it was possible to catch a ride back to the nearest town for "a little hell raisin'"; but Heaven help anyone who failed to show up ready for work at sunrise Monday.

**Figure 1-13** Typical railroad survey crew. *Courtesy of Dale Landon Thomas*

The first Portland Cement concrete paved roads in the United States were built around the courthouse in Bellefontaine, Ohio, in 1891. This year also saw the production of the first automobile and the establishment of the Office of Road Inquiry by the Department of Agriculture. The government was now in the highway business.

The railroads, now the Northern Pacific, Southern Pacific, and Santa Fe, were also in operation and needed feeder roads to connect with their stations.

Roads were built better and farther. In 1903, Dr. A. Nelson Jackson drove a Winton motorcar across the United States to become the first transcontinental tourist. By 1904, there were 153,662 miles of paved roads. By 1914, there were seven cars to the mile of paved roads. In 1916, President Wilson signed the Federal Aid Road Act. By World War II, there were 1,384,000 miles of surfaced roads in the United States.

The growth of the System of Interstate Highways provided many jobs for survey technicians. These technicians were employed by the state highway departments after the federal road programs tapered off in the early 1950s. With the number of cities growing all over the country, there became a need for a specialized type of surveyor – the city surveyor.

A survey textbook of 1886 recommends a city lot to be 20- to 30-feet wide and 100- to 180-feet long to provide for a "high-class residence" with "room for a large lawn." Streets are to "vary from, say, 210 feet, (the width of Ocean Parkway in Brooklyn) to 40 feet for some unimportant short side street." Residential streets were to be 50- to 80-feet and main thoroughfares not less than 100-feet across. These widths were necessary to allow room to turn a team and wagon around without having to go to an intersection.

The Act of July 1, 1862, gave the railroads a 200-feet right-of-way and all the odd numbered sections for 10 miles on each side of the tracks. This caused many townships along the route to be hastily surveyed and resulted in many partial townships requiring resurveys as they were sold and resettled.

On March 30, 1867, Alaska was purchased from Russia adding 590,066 square miles to the United States.

The Desert Land Act of March 3, 1877, allowed homesteading of up to 640 acres of irrigable desert if the settler could "prove up" by bringing in water and farming at least a portion of the claim. This opened previously excluded land that had been thought to be worthless for surveying and settlement.

A provision was made in 1879 that allowed settlers to make deposits for surveys of unsurveyed public land. In the West, this act unwittingly led to the "Benson syndicate frauds." John Benson was a deputy surveyor in California who hired deputy surveyors, clerks, and draftsmen. He paid off a few clerks in the surveyor general's office, then filed applications for deposit surveys in remote areas. The "deputy surveyors" were then "hired" to make the surveys. These were usually only "paper surveys"; no actual surveying work was done, and the notes and plats were faked. If some of the land was actually occupied, some crude surveying work was performed to satisfy the "locals." The surveys were then filed and paid for at the highest rates allowed by law. Many thousands of miles of fictitious surveys were paid for with government funds.

This system operated until 1885 when Congress began to get tight with money, allowing only $300,000 for all surveys and taking the profit out of Benson's scheme.

## THE UNITED STATES GEOLOGICAL SURVEY

The Civil Appropriations Act of March 3, 1879, created the United States Geological Survey (USGS) under the Department of the Interior. The USGS was made responsible for classifying the public lands and making some of the rectangular surveys.

In 1881, John B. Marston, the county surveyor for Toledo, Ohio, performed the first rectangular survey using a **steel tape** instead of a **chain**.

George W. Garside, a deputy surveyor from Nevada, surveyed the Aurora Lode Mineral Survey No. 41 on August 9, 1887. Located near Juneau, Alaska, this was the first approved official survey in Alaska.

The 1894 Manual of Surveying Instructions issued June 30, 1894 eliminated the use of the **magnetic compass** for surveying public lands. In October of 1896, the first circular, titled "Restoration of **Lost** or **Obliterated Corners** and Subdivision of Sections," was issued.

On April 17, 1905, the stone monument for the initial point of the Copper River meridian was set in Alaska by Alfred B. Lewis, deputy surveyor. The first rectangular surveys in Alaska were approved by the Surveyor General for Alaska, William L. Distin, on January 28, 1908.

The contract system of surveying the public land ended on June 25, 1910, with the Civil Appropriations Act, which provided, "The surveys and resurveys to be made by such competent surveyors as the Secretary of the Interior may select, at such compensation not exceeding two hundred dollars per month. . . ." The first survey under the new system was T 14 S, R 100 W, Sixth Principal Meridian made by A. C. Horton Jr., U.S. Surveyor, and Jay P. Hester, U.S. Transitman, over a period of two and one-half months from July through September, 1910.

The Act of June 25, 1910, provided $100,000 to survey the public lands in Alaska. R. H. Sargent was appointed supervisor of a large survey party that established the Fairbanks base and meridian, working under instructions from USGS. This was probably the last rectangular survey made by the Geological Survey.

## THE BUREAU OF LAND MANAGEMENT (BLM)

On August 24, 1912, Alaska became a United States Territory. The Act of July 16, 1946, put the GLO, the Grazing Service, and other agencies all under the title of the Bureau of Land Management (BLM). The GLO was abolished and the BLM assumed the duties of surveying all the public lands left unsurveyed.

# Review Questions

1. In what year did the English Statute of Frauds become law?
2. How does the Statute of Frauds affect modern land titles?
3. Define the term "Free Socage."
4. Who were Mason and Dixon?
5. Who was Andrew Ellicott?
6. Who was Benjamin Banneker?
7. Who was David Rittenhouse?
8. What was the first state created out of the Northwest Territories?
9. From whom did the United States purchase the Province of Louisiana (the Louisiana Purchase)?
10. Who did President Jefferson pick to survey the Louisiana Purchase?
11. For what is Lt. Zebulon Pike famous?
12. Name a famous "mountain man" known for exploring the West.
13. Who made the map that opened the Oregon Territory to settlement?
14. For what is John C. Fremont known?
15. Which group of men were responsible for surveying and mapping the West in the 1800s?
16. Name three of the officers who performed the surveys of the West.
17. Where did the Union Pacific and Central Pacific Railroads meet?
18. In what year was the Homestead Act signed?
19. In what year was the U.S. Geological Survey organized?
20. Name three of the four famous surveyors of the U.S. Geological Survey.
21. Who was responsible for dividing the public lands into townships and sections?
22. In what year were the first Instructions for the Survey of the Public Lands issued and by whom?
23. In what year was the Burt's solar compass brought into general use?
24. In what year was the first *Manual of Instructions for Subdivision of the Public Lands* published?
25. When was the first concrete paved road built and where?
26. When was the first motor car driven across the United States?
27. In what year was use of the magnetic compass eliminated in the surveying of public lands?
28. When did the contract system of surveying public lands end?
29. In what year did the Bureau of Land Management assume the duties of surveying the public lands?

30. In your own words, why is the history of surveying important to a survey technician?

31. Name a well-known local surveyor of the past century and describe one of the surveys made by that person.

# Surveying Equipment and How to Use It

2

## Objectives

After completing this chapter, the student should be able to:

1. Identify the different types of equipment used in the surveying industry.
2. Specify the uses for each type of survey equipment identified.
3. Demonstrate the proper handling of each type of survey equipment available to the student.
4. Demonstrate the proper care of equipment at the end of the project or day.

In this chapter, you will acquire knowledge of the various types of survey equipment in current use. Where possible, there will be a short history of the piece and instruction on its common use. Please be aware that there are nearly as many ways to use survey gear as there are surveyors to use it, so all instructions are merely suggestions. You will develop your own unique method to get the job done as you gain proficiency. One of the purposes of this book is to help you avoid forming bad habits; but from then on you are on your own.

One of the oldest items of survey equipment still in use is the **plumb bob**. The Egyptians are known to have used a plumb bob in building and a plumb line for leveling (see Figure 2-1). They also had an instrument called a merchet that used a plumb line sighted through a slit palm leaf for alignment. Over the centuries

**THE PLUMB BOB**

**Figure 2-1** Egyptian level. *Courtesy of Cosette Landon Sutter*

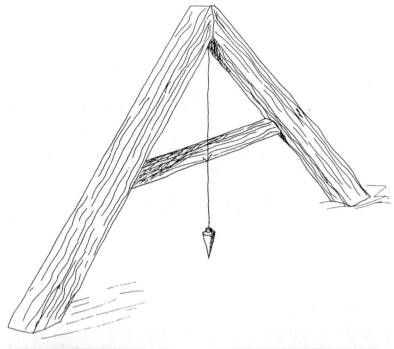

**Figure 2-2** Plumb bob

the plumb bob has taken on many shapes, although most at least resembled the one we use today (see Figure 2-2).

Plumb bobs come in several weights, the most common being the 16 and 18 oz. for **chaining**. If you are working in windy conditions, a heavier bob (24 to 36 oz.) will swing less, but it can get very heavy when you lug it around all day on your hip.

A handy way to control the string, which should be as long as you can reach with your arms extended above your head, is the **gammon reel**. If you do not have a reel, the string can be carried looped across your shoulders. You must be careful; if the string catches on something, it can give a bad rope burn across the back of the neck. When you use the bob, the tip can be steadied by gently tapping it on the point. Brace your hand and arm as if you were holding a camera to give sight (steady the bob for use as a good target) as in Figure 2-3.

Be sure to look up at the instrument technician occasionally to see if any hand signals have been given. (See Appendix A for Surveyor's Hand Signals.) When the point is directly over the mark, call out, "Good, good . . ." in a loud voice. The use of the plumb bob in taping is covered in Chapter 3 and its use on instruments in Chapter 4.

If the point breaks off leaving the screw-end in the bob, it can sometimes be worked out with the tip of a concrete nail or a knife. If that fails, carefully cut a slot in the tip of the bob with a hacksaw just deep enough to slot the broken tip, which can then be backed out with a screwdriver. If you carry a piece of hacksaw blade in your chaining pouch, the next time you break a tip just slot the broken tip and screw it out.

Be careful with the plumb bob's sharp point. If you are not wearing sturdy boots, a bob point stuck in your foot can be a painful experience. Always carry the plumb bob snapped into its sheath when not in use; it may prevent an accident or, at least, will prevent the loss of an essential tool.

**Figure 2-3** Bobbing over a point

## THE LOCKE HAND LEVEL

The Locke **hand level** has been in use since the early 1800s. It can be used to find a level for taping or for rough leveling on side shots in **topographic** surveys. The hand level can also be used to transfer grades over short distances when properly braced. Some models have **stadia** hairs and magnifying lens for those of us who need extra visual aid (see Figure 2-4).

The hand level is also known as a **"P gun."** This name is probably derived from its use on "P lines" or preliminary survey lines. In use, the hand level is held to the users eye and sighted on the object to be leveled (see Figure 2-5).

On one side of the field of vision, you will see a **mirror** with a

**Figure 2-4** Hand levels

**Figure 2-5** Using plumb bob and hand level

**bubble** in front of it and a cross hair across it. When the bubble and the cross hair are even with each other, whatever coincides with the cross hair in the field of vision is level with the hand level.

Now that you have your plumb bob and hand level, you will need a way to carry them. A good, wide, heavy duty belt and a leather carpenter's pouch will serve admirably (see Figure 2-6). Get a pouch with a couple of pockets and a place to carry a 10- or 12-ft. carpenters tape or "yo-yo," graduated in inches and tenths of feet. Get a 1-inch-wide tape if you can; it makes a good substitute rod. You'll also need a pocket for your **keel** or lumber crayon, which is used to mark on the guard stakes or **lath**. A small, soft rubber ball on a length of leather lace or "wang" to put your "cup-tacks" in will save your fingers. Reaching into a pouch to get a sharp tack can be a painful experience, so be careful with sharp objects!

One more thing you will need is a sharp knife to dig out splinters, sharpen lath, and for hundreds of other uses. One with a sheep-foot blade is great for splinters. The most important piece of gear you will need, however, is a good pair of leather boots. The U.S. Forest Service requires at least an 8-in. top and Vibram soles. Your feet are going to have to carry you for a lifetime so take good care of them. Do not buy cheap boots or socks.

**Figure 2-6** Surveyor's taping belt

## RODS AND CHAINS

Ancient people needed a way to describe distance. The day's journey was probably the first way to measure how far to travel from here to there. More sophisticated measurements were needed as people gathered together and settled in villages. The pace was a good method to describe a field but, as the land was divided, an even smaller unit was necessary.

A common unit of measure in ancient Egypt was the cubit, the distance from the elbow to the fingertips. It was inconvenient to measure fields with the forearm, however, so sticks of wood or rods were used for short distances.

As land was claimed by owners who wanted to raise crops, it became necessary to mark the boundaries with **monuments**. This worked out until the Nile flooded and wiped out the landmarks. Priests, or Harpedonaptae, decided it was necessary to be able to remeasure the distances with greater **accuracy**. They had rope to do this but rope stretches when pulled. Heron of Alexandria describes how they overcame this problem. "The rope is stretched between two stakes and, after it had stayed in this position for a considerable length of time, it is stretched again. When the process has been repeated, the rope is rubbed with a mixture of beeswax and rosin. It is better, however, instead of stretching the rope between two stakes, to suspend it vertically for a long time with a

**Figure 2-7** Gunter's chain

sufficiently heavy weight attached to it." This system produced a rope that was accurate to 1 ft. in 2000.

For more accurate measurement, **rods** with bronze caps were laid end to end. The English rod of 16 ft. 6 in. was the standard of measurement into the early 1900s.

Edmund Gunter, an English mathematician, who invented the *66*-ft. surveyors chain in the early 1600s, also introduced the cosine and cotangent to trigonometry. **Gunter's chain** was used to measure all the public lands in the United States (see Figure 2-7).

## THE STEEL TAPE

With modern metallurgy the steel tape, or **engineer's tape,** came into general use in the twentiety century (see Figure 2-8). The engineers tape is graduated into feet, tenths of feet, and hundredths of feet or in meters. The common lengths are 100, 200, 300 feet,

**Figure 2-8** Engineer's tape

**Super Hi-Way Chrome Clad**®

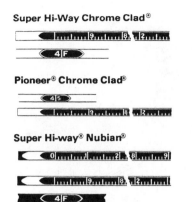

**Pioneer**® **Chrome Clad**®

**Super Hi-way**® **Nubian**®

**Engineer's long tapes: blade styles
E1 Chrome Clad** ®
**and yellow clad E1 – E6**

**E2 Chrome Clad**

**Figure 2-9** Steel tapes. *Courtesy of the Lufkin Division of Cooper Tools*

and also 30 meters. The steel tape (Figure 2-9) is still used for distances of 200 ft. or less, whereas the **Electronic Distance Measurement (EDM)** is most often used for longer distances or to measure across highways and obstacles.

The EDM first came into use in 1946 when Dr. Bergstrand developed the geodimeter in Sweden. The geodimeter used a modulated light-wave pulse reflected off a mirror. In 1956, the tellurometer was invented by T. L. Wadley in South Africa. The tellurometer used microwave pulses instead of light. This allowed the microwaves to travel through canvas (a benefit that saved the author from freezing on the trans-Alaska pipeline). The Hewlett-Packard 3800 came out in 1970 (see Figure 2-10). The 3800 fits on the **tribrach** mount in place of the theodolite, greatly simplifying the measurement.

**ELECTRONIC DISTANCE MEASUREMENT (EDM)**

**Figure 2-10** HP 3800. *Courtesy of Hewlett-Packard Company*

**Figure 2-11** A) Zeiss Elta 2, front view; B) Zeiss Elta 2, back view. *Courtesy of Carl Zeiss, Oberkochen, Germany*

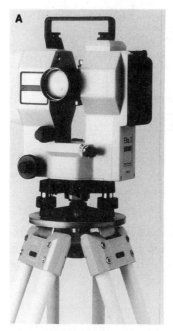

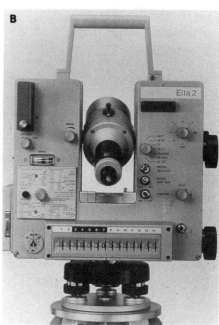

In 1968, The Carl Zeiss Company introduced the REG ELTA 14 as the first electronic recording tacheometer, combining the theodolite and EDM in one instrument. In 1978, Zeiss introduced the **total station,** the Elta 2 (see Figure 2-11).

The beam of light from the light wave system is returned by a **prism** or mirror (see Figure 2-12). Some surveyors use a small mirror mounted on a rod or hung on a plumb bob string (see Figure 2-13). This small mirror can be placed on the **foresight** or point to be set by the survey **technician,** and the distance can then be observed by the instrument operator (see Figure 2-14). Because of the higher cost of the microprism compared with a plumb bob, the technician must take care not to drop or lose the prism when working through heavy brush.

For rough work or when using a steel tape is not practical, a woven or cloth tape can be used (see Figure 2-15). Some of these woven tapes have metallic strands woven into them for strength and to prevent stretching. **NEVER** use metallic tapes near power lines or other sources of electric current.

A **range pole** can be used for longer distance sighting. Range poles are red-and-white striped wood- or plastic-coated metal and are either hand held or propped up in a tripod over a point (see Figure 2-16). Once again, BE CAREFUL WHEN WORKING AROUND ELECTRIC LINES with metal poles! Watch the overhead clearance and stay well away from all high tension power lines.

**Figure 2-12** Prism on tripod

**Figure 2-13** **(Top left)** Prism with plumb bob over a point

**Figure 2-14** **(Top right)** Prism on a rod

**Figure 2-15** **(Bottom left)** Cloth tape

**Figure 2-16** **(Bottom right)** Range pole in a tripod

**Figure 2-17** Using a right angle prism

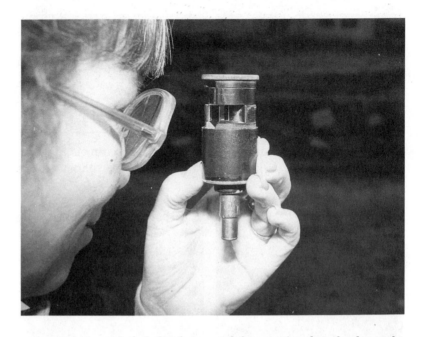

**Figure 2-18** Using a metal detector

A machete or brush hook is used for cutting brush along the line of sight. Use an axe or chainsaw for heavier brush and trees. All edged tools are safer when kept sharp. When you use a file to sharpen a tool, always hold the file handle to protect your hands from the sharp edge. When you use a cutting tool, always be aware of where the rest of the crew is at all times and never cut toward another person. Do not cut toward yourself when you swing an axe or machete. The branch might be soft or hollow and the follow-through of the cut might cut you badly. Speaking of getting cut, do you have an up-to-date first aid card? Everyone on the crew should have a current Red Cross industrial first aid card. Be sure you take along the first aid kit too.

Use a **right angle prism** for finding an approximate perpendicular to your line of sight. If you get a double right angle prism it can also be used to get approximately on line between two points (see Figure 2-17).

A metal detector can be used for finding buried pipes and pins. Some commercial survey markers have a magnet built in to make the markers easier to find with a detector. Some detectors have a small tip instead of a large pan head that allows them to be poked into smaller places and under bushes (see Figure 2-18).

A good set of two-way radios makes long-distance communication easier, especially on a noisy job site. Do not use two-way radios near a site where blasting is being done. The energy from a broadcast might set off an electric blasting cap and cause an explosion. Hand signals can be used at moderate distances or when the noise level is too high for normal conversation. Be sure to learn the hand signals used in your area as they vary from state to state and company to company. It is particularly important to learn the proper hand signals when working around heavy equipment; they could save a person's life or prevent an accident.

## THE LEVEL AND ROD

The basic tool for measuring differences in elevation is the level and rod (see Figure 2-19). As an entry level survey technician, you will need to become familiar with the different types of leveling rods. A standard rod is made of hardwood with a metal face. The top and bottom are protected by metal caps, which should be checked for looseness and tightened before they are used. A **Frisco** rod is a three-piece rod. A **Philly** rod has two pieces. Lenker developed the direct reading rod, which is known as a **Lenker rod**, regardless of the manufacturer.

**Telescoping** rods are available in lengths up to 25 ft. or more when longer rods are required (see Figure 2-20). Once again, BE CAREFUL WITH OVERHEAD WIRES! For **stadia shots** a folding rod graduated in feet, tenths, and half tenths is easy to read at longer distances (see Figure 2-21). For precise level runs, a single-piece rod with a face of **Invar**, a low expansion metal, is used. Invar rods come in matched pairs and are very expensive.

A rod level is handy for keeping a leveling rod or range pole plumb. The rod level is designed to be held against the rod while a circular, or **fish-eye**, bubble is centered for plumb. The rod level can be adjusted for level by tightening the three adjusting screws under the lip. Use a good carpenter's level and check both ways for plumb.

The level is the simplest of the instruments and one of the oldest. The Egyptians used a level made from two sticks of wood fastened together in the form of an "A" with a plumb bob hung from the apex of the A. When the plumb line bisected the cross arm of the A, the legs were level (see Figure 2-1). Heron of Alexandria developed the water level around 100 B.C. Heron's level was a wooden block, about 6 ft. long, on which a long bronze tube was mounted. On each end of the tube, glass tubes were mounted at right angles. When the tubes were filled with water, the operator could sight across them at the rod for level.

The **spirit level** was first mentioned in 1666 by Thevenot in France. Development of the spirit level was slow because of the difficulty of getting the curvature in the glass tube to give the bubble a smooth movement. The wye level of the 1800s (see Figure 2-22) made use of the spirit level. A frame containing the bubble

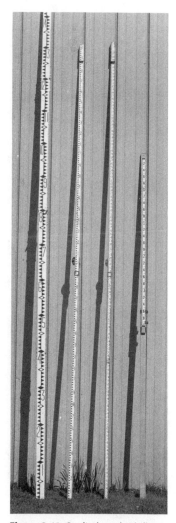

**Figure 2-19** Stadia board, Philly rod, Lenker rod

tube was mounted on a tripod and the telescope with the cross hairs was mounted above it on the wyes. To check for level, the telescope was lifted out of the wyes and reversed.

The next modification was the **dumpy** or **engineers** level which is similar to the wye except the telescope is fastened to the base (see Figure 2-23). The dumpy level is still used for work on construction sites where vibration from the heavy equipment causes undesirable vibration in the compensator of an automatic level.

With the **tilting** or **spirit bubble** level, both ends of the bubble can be seen in the field of view of the telescope. Bringing the two ends into coincidence by means of an adjusting screw allows more accurate leveling of a line of sight.

**Automatic** or **self-leveling levels** make use of a compensator to make the line of sight level, even if the instrument is slightly out of level. The level is rough leveled first with a fish-eye bubble. When the bubble is within the center circle the compensator maintains a level line of sight when the instrument is rotated (see Figure 2-24). This feature is useful and makes the auto-level faster to use than a dumpy. One problem is the possibility of the compensator sticking. To check for a stuck compensator, gently tap the tripod leg as you watch the cross hair through the telescope. The cross hair will appear to drift up and down in the line of sight. If the cross hair returns to its original reading on the rod, the compensator is functioning properly and is free to move. Newer auto-levels have a magnetically dampened compensator and a

**Figure 2-20** Telescoping rod

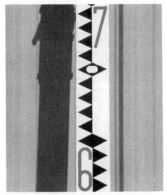

**Figure 2-21** Stadia board

**Figure 2-22 (Right)** Wye level

**Figure 2-23** Dumpy level

**Figure 2-24** Zeiss auto-level

more ruggedly constructed mounting system. On older auto-levels, care must be taken when carrying the level to prevent the pendulum from swinging back and forth, wearing out the mounts. A good rule of thumb is to carry the level in the same position it is mounted in the case or, for long distances, return it to the case.

## MEASURING ANGLES

The earliest known instrument for measuring angles was probably a square. A tomb painting from about 1450 B.C. shows a square similar to the carpenter's square that is used today. The Babylonians used the 3-4-5 triangle as early as 1800 B.C. to lay out a right angle. If two sticks are fastened together at right angles and plumb bobs are fastened to the outside ends of the sticks, a person could sight across the plumb strings and lay out a 90° angle.

The Romans had a similar instrument called a **groma** (see Figure 1-2). The Roman groma was mounted on a staff with an arm on top to allow the center to be plumbed over a point. The bobs may have been suspended in cups of water or oil to dampen the swing in the wind. The same form of instrument has also been known as the "Grecian star" and the "surveyors cross." The Greek mathematician, Heron, invented an instrument known as the "dioptra" that could be used as a level or a right angle instrument by changing the upper section.

The earliest known instrument for measuring vertical angles was the **astrolabe**, which dates to about A.D. 100. The astrolabe was a brass circle that was divided into 360°. Fastened to the back by a pin through the center of the circle was a sighting bar or **alidade**. In use, the astrolabe was suspended from the thumb by a string and the object was sighted through the alidade sights. The angle could then be read from the circle (see Figure 2-25).

**Figure 2-25** Astrolabe

The "polimetrum" appears in a 1512 edition of a book on map making and surveying by Martin Waldseemuller. Having both horizontal and vertical circles, the polimetrum was the earliest European prototype of the theodolite and **transit**. In Thomas Digges's book *Pantometria*, published in 1571, he described an instrument called the "theodolitus" or "instrument topographical." This was the origin of the modern theodolite in England.

In 1631, Pierre Vernier invented the auxiliary circle that allowed the circles to be read to the nearest minute of arc. The English astronomer, Gascoigne, put cross hairs in the telescope in 1640. Geminiano Montanari used stadia hairs to measure long distances in 1674. The hairs are set at a ratio of 1:100 so that 1 ft. read on the rod equals 100 ft. of horizontal distance on the ground. With these refinements, we now have the basic modern transit and theodolite.

The Carl Zeiss Company of Oberkochen, West Germany, has been a leader in the manufacture of survey instruments since the 1800s. In 1899, Zeiss introduced the stereo range finder, followed by the internal focusing instrument in 1908. Since then, Zeiss has introduced a number of "firsts." From 1921 through 1931, Zeiss's contributions included a measuring theodolite, optical theodolite, optical scale reading theodolite and the first modern one-second theodolite (see Figures 2-26 and 2-27).

In 1950, Zeiss introduced the first automatic level, the Ni2. In 1968, the first electronic recording tachometer, the Reg Elta 14, was marketed, followed by the first lightweight, compact EDM/

**Figure 2-26 (Left)** Optical theodolite. *Courtesy of Carl Zeiss, Oberkochen, Germany*

**Figure 2-27 (Right)** One-second theodolite. *Courtesy of Carl Zeiss, Oberkochen, Germany*

**Figure 2-28** Zeiss Elta 4. *Courtesy of Carl Zeiss, Oberkochen, Germany*

theodolite in 1976. The first completely computerized surveying system, the Elta 2, and the first lightweight electronic tacheometer, the Elta 4, helped revolutionize surveying in 1978 (see Figures 2-11 and 2-28). More recent developments include the Prog 12 program module for the Elta 2. It is the first software program providing twelve surveying routines for direct, in-the-field computation.

The transit has been in use since the seventeenth century by surveyors and civil engineers (see Figure 2-29). It is regarded as the universal surveying instrument because of its many uses. What we call the transit in the United States is actually the **vernier** theodolite. There are optical reading transits but they are more widely known as theodolites in this country.

The transit has three main parts: the leveling head, the lower plate or motion, and the upper plate or motion (see Figure 2-30). The upper and lower motions contain the verniers and circles that are used to read the horizontal and vertical angles. The leveling head of the American transit has four leveling screws; the European transit has three leveling screws. The leveling head can be shifted on the footplate to allow precise centering over the point.

**Figure 2-29** Transit

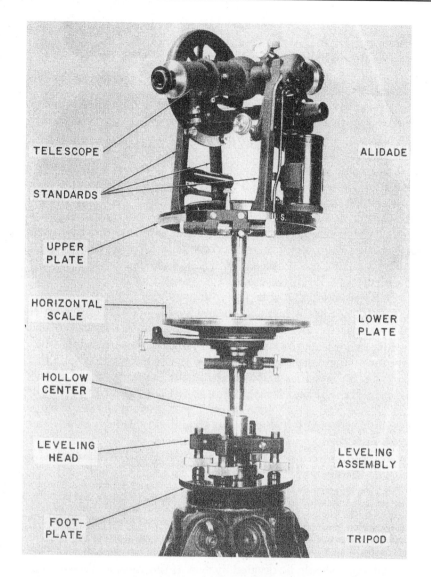

TELESCOPE

STANDARDS

UPPER PLATE

HORIZONTAL SCALE

HOLLOW CENTER

LEVELING HEAD

FOOT-PLATE

ALIDADE

LOWER PLATE

LEVELING ASSEMBLY

TRIPOD

**Figure 2-30** Exploded view of a transit. *Courtesy of the Department of the Army Manual, 1964*

Some transits have an **optical plummet** for centering but most have a metal hook to which the plumb bob string is attached by a slipknot or looping around the gammon reel. The leveling screws are turned until the plate level vials are centered. The rule is: thumbs together, thumbs apart. This action will move the bubble in the direction of the left thumb. The alidade, or **upper motion**, is then rotated 180° and the bubbles again checked for center. If a small difference is found, split the difference with the leveling screws and repeat the adjustment. Once the bubbles remain in the same relative position when rotated through 180°, the transit is level.

The next step is to release both the upper and lower motion lock screws and set the "A" vernier to zero. This can be done by placing the thumbs on the upper motion and using the index fingers to rotate the lower motion. Once the zero mark is close, lock the upper motion-lock screw and fine adjust the vernier with the upper motion-**tangent** screw. It is sometimes easier to line up the marks on each side of the zero mark than to try to line up the zero line on the upper and lower plates. With zero locked into the upper motion, and the lower motion loose, rotate the alidade until the scope points at the **backsight**. The scope can be rotated vertically by loosening the vertical motion clamp screw.

Once the scope is aligned on the backsight and focused by turning the focus knob located on top of the scope, move your eye slightly from side to side. If the cross hairs appear to move on the target, you have **parallax** in the cross hairs. Adjust the parallax out by focusing the scope on infinity and looking toward the sky or a blank wall. Focus the cross hairs by rotating the eyepiece until the hairs are crisp and sharp to your eye. Now focus on the backsight with the scope-focus knob and lock the lower motion with the lower motion clamp. Check again for parallax and fine adjust the alignment with the lower motion tangent. Now release the upper motion lock and rotate the scope to the foresight. Focus the scope on the foresight and lock the upper motion clamp. Use the upper motion-tangent screw to fine adjust the alignment. Always turn the tangent screws into the spring to prevent the scope from moving accidentally if the motion should bind. Once you are satisfied with the alignment, read the angle on the "A" scale. Read the degrees on the outer scale and the minutes and seconds on the vernier. Remember, "turn to the right, read to the left" and vice versa (see Figure 2-31).

To repeat or "double" the angle, loosen the lower motion, invert the scope, resight the backsight with the scope inverted (the focus knob will be on the bottom), and lock the lower motion. Repeat the procedure for turning the angle and read the angle again. This second angle should be divided by two. If the mean angle is within a few seconds of the first angle, the angle is good. Repeat this procedure for several sights around the horizon. The mean angles should add up to 360° if you have done everything right.

Reading vertical angles is similar to reading **horizontal angles** except that you will be reading "plus or minus angles" instead of right and left. Release the vertical motion clamp and set the center cross hair on the target. Notice the other two hairs in the scope. These are stadia hairs, which are used for measuring distance. Be careful to get the center hair when reading vertical angles, especially if you wear glasses that restrict your field of vision through the scope. With the center hair rough set on the target, fine adjust

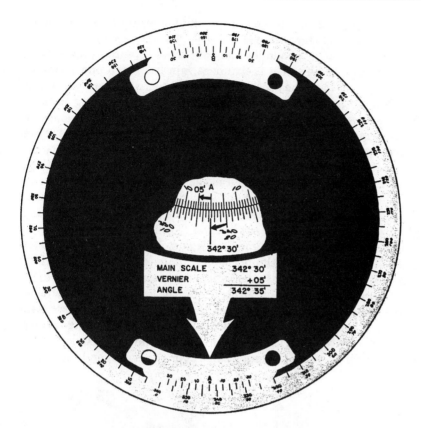

**Figure 2-31** Vernier scale. *Courtesy of the Department of the Army Manual, 1964*

| MAIN SCALE | 342° 30' |
| VERNIER | +05' |
| ANGLE | 342° 35' |

using the vertical motion-tangent screw. Move around to the side of the transit and read the direct vertical angle. Notice this time the vernier is on the outside of the circle. Record the angle and reverse the upper motion or alidade and repeat the process with the scope inverted. Take the mean of the two vertical angles. The difference is called the vertical index of the instrument.

**THE THEODOLITE**

Setting up the theodolite is slightly different than setting up the transit (see Figure 2-32). First, be sure that the tribrach release is locked before moving the instrument. Set the instrument over the point and set one leg of the tripod firmly in the ground by pushing the top. Then look through the optical plummet to see if the point on the ground is visible. If you cannot see the point, place the toe of your boot next to the point to give yourself a bigger target and better direction. Pick up the two unset legs with both hands and, while still looking through the optical plummet, rotate the tripod head until the cross hairs of the plummet are approximately centered on the point. Try to keep the plate as level as possible so you won't run out of screw when you level the instrument. Now, set

**Figure 2-32** Setting up a theodolite

**Figure 2-33** Leveling a three-screw theodolite

the two legs firmly in the ground and make sure that the point is still visible in the plummet.

Notice that a theodolite has only three leveling screws. When leveling a three-screw instrument, always leave one screw untouched and use the other two to level the instrument (see Figure 2-33). Center the plummet cross hairs on the point using the leveling screws. Center the fish-eye bubble by adjusting the length of the tripod legs. Raise the leg that is opposite the bubble or lower the leg that is nearest the bubble. When the bubble is close to center, the legs can be stepped on to fine adjust the bubble. Check the plummet cross hairs to be sure that they are still on point. If the hairs are slightly off the point, loosen the tripod mounting screw and slide the tribrach over the point. Be sure not to rotate it as it is being moved or it will not be level. Tighten the tripod mounting screw, check for center, and level again. Now, using only two leveling screws, adjust the horizontal plate-leveling bubble. If the alidade is turned so that the bubble tube is aligned with one of the leveling screws, it will be easier to get the bubble centered. Rotate the alidade 90° so that the bubble tube is now aligned with the other screw and center the bubble. Check to be sure the cross hairs are still centered on the point. If the hairs have moved, loosen the tripod screw, slide the cross hairs over the point, and relevel the

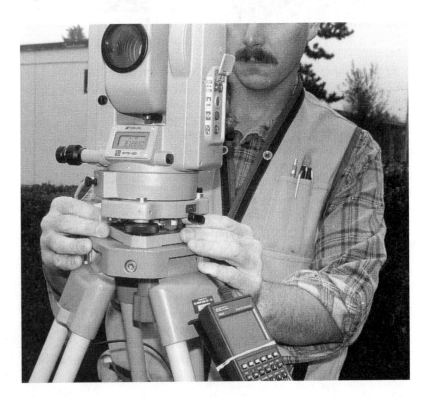

plate bubble. Repeat until the cross hairs are centered on the point and the plate bubble stays centered when the alidade is rotated through a full circle.

There are two types of theodolites: the **repeating** theodolite and the **directional** theodolite. There are many brands of theodolites, however, and there seem to be more every year. Because there are so many different models, it would be impossible to describe all of them here; therefore, one common model of the repeating type that is most often used by surveyors will be described here.

Repeating theodolites are preferred for most everyday work because angles can be turned in either direction and they can be set to zero. Directional theodolites do not have a lower motion and are preferred for control surveys because they are usually more precise.

The lower motion on a repeating theodolite is similar to a transit, having a lower motion-lock clamp and a tangent screw. Some models have a circle-locking lever so be sure to read the owners manual, unless someone is willing to show you how to operate the type of theodolite you will be using.

The upper motion is the same as the transit and is used the same way. The major difference you will notice is that there are two eyepieces: a large one and a smaller one to the right of the larger one (see Figure 2-34). The larger one is the telescope eyepiece, and it is focused and the cross hairs adjusted in the same way as the transit. The smaller eyepiece is used to read the horizontal and vertical circles.

On the side of the alidade is a small mirror. When you look through the reading scope (the smaller one), adjust the mirror to reflect the light on the reading circles. If the circle markings are not clear, focus the eyepiece until the markings are clear and sharp.

On some theodolites, an optical micrometer is used to set the minute and second readings. On this type of instrument, it is important to first set the micrometer to zero using the micrometer-adjusting knob on the side of the alidade. Next, the horizontal circle can be set to zero, using the upper and lower clamps and upper motion tangent, as with the transit (see Figure 2-35).

If you turn the micrometer-adjusting knob as you look in the reading microscope, you will notice that both the horizontal and vertical circle verniers move, but the alidade itself remains on the sight. This is because the micrometer is an optical adjustment only and does not move the instrument. To read a vertical angle, set the center cross hair of the telescope on the sight and lock the vertical motion-lock screw. Fine adjust, using the vertical motion-tangent screw, then bring the degree index into line, using the micrometer. Finally, read the minutes and seconds on the auxiliary scale. The horizontal angles are read the same way.

**Figure 2-34** Directional theodolite showing two eyepieces. *Courtesy of the Department of the Army Manual, 1964*

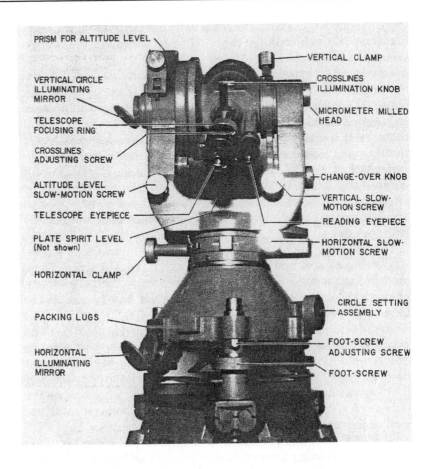

**Figure 2-35** Theodolite optical micrometer circle. *Courtesy of the Department of the Army Manual, 1964*

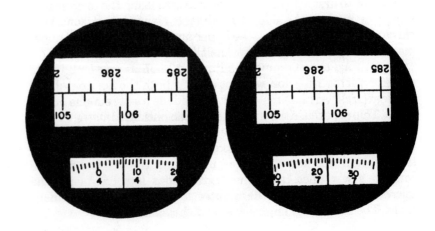

To lay off an angle, reverse the process. First, set the minutes and seconds on the auxiliary scale and then the degrees on the horizontal or vertical scale. The cross hairs will now be on the required point.

A directional theodolite is similar except that you cannot set the zero. To read an angle with a directional instrument, read the circle on the backsight and record the angle. Next, turn to the foresight, using the upper motion lock and tangent. Again, read the circle and record the angle. Now subtract the backsight reading from the foresight reading. The result is the angle between two points.

The latest improvement in theodolites is the electronic digital theodolite (EDT) that has an LED (light emitting diode) or LCD (liquid crystal display). It reads to the nearest second and can be set to zero with the touch of a button.

On most newer electronic theodolites it is possible to interface a **data collector** to the instrument (see Figure 2-36). The data collector records the readings as they are taken and can download directly to a computer. If an EDM is added to the electronic digital theo-

## THE TOTAL STATION

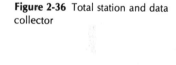

**Figure 2-36** Total station and data collector

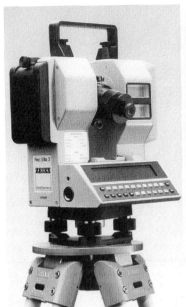

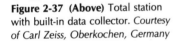

**Figure 2-37 (Above)** Total station with built-in data collector. *Courtesy of Carl Zeiss, Oberkochen, Germany*

**Figure 2-38 (Top right)** Total station: A) front view; B) back view

dolite and a computer is added to reduce raw data to horizontal distances and corrected angles, the expression "total station" takes on real meaning. Several of the total stations with built-in computers will give state plane coordinates and elevations for points in the field (see Figure 2-37). State plane coordinates is a system of coordinate grids, each state having one or more zones, which was designed by the U.S. Coast and Geodetic Survey for state coverage. They are also able to do automatic radial layout from the same coordinates (see Figure 2-38).

## SURVEYING WITH GPS (GLOBAL POSITIONING SYSTEM)

### "Surveying with GPS"

Global Position System (GPS) satellite surveying is a 3-dimensional measurement system based on observations of the radio signals of the Department of Defense's NAVSTAR (NAVigation Satellite Time And Ranging) GPS system. The observations are processed to determine station positions in an Earth-centered Cartesian coordinate system (X, Y, Z), which can be converted to geodetic coordinates (latitude, longitude, and height) on the reference ellipsoid. The coordinate system of the satellites is the World Geodetic System of 1984 (NAD 83). With adequate corrections of the geoid, orthometric heights (height above sea level) can be computed for points with unknown elevations.

Block II satellites, with 21 to 24 satellites, were planned for full operation by 1991, but delayed because of the Challenger accident. They will be deployed in 6 orbital planes inclined 55 degrees to the equator. Both Block I and Block II satellites will be in a near-circular orbit, at an altitude of approximately 20,000 kilometers (12,000 miles) and an orbital period of 12 hours. The completed complement of Block I and Block II satellites will allow for continuous 24-hour satellite observations.

GPS is a one-way ranging system: the measuring wave is transmitted by the satellite as it moves through space and the GPS receiver with its antenna on the ground is a radio receiver. A very accurate atomic clock on the satellite sets, or synchronizes, a quartz clock in the receiver. Software in the GPS receiver assigns time tags to the date, and computer software corrects for clock errors and phase ambiguities.

GPS provides an excellent means for measuring distance. Two GPS receivers placed on two points (one known or both unknown), observing the same satellites at the same time, will give accurate distance measurement, regardless of terrain or distance between stations.

In the static (fixed) mode, with two or more GPS units receiving signals from the same satellites at the same time, one GPS unit is always positioned on a known point (NAD 83 or WGS 84 coordinates); the other unit(s) on the unknown point(s). This observing period is called a session. When the unknown point(s) in observing Session 1 become known, one unit remains on a known point while the other(s) are "leap-frogged" to other unknown point(s) for Session 2, etc. The observations are processed to obtain the components of the baseline vector between observing stations (dX, dY, dZ). The coordinate differences between the receiver at the known point and the receiver at the unknown point(s) can be determined to an accuracy of 1:100,000 or better. A minimum of four satellites should be visible at the same time to do 3-dimensional measurements (three will work for horizontal control only).

In the static mode, the number of points that can be established in a day is determined by the number of receivers used. The greater the number of receivers, the greater the productivity; for example, 2 units give one new point per session; 3 units, two new points; 4 units, three new points; etc. Observing time can be minimized by scheduling sessions when the satellites' positions are the greatest. In general, a minimum of 45-minutes observing time is required when using the static mode. With efficient planning, as many as 4 sessions can be completed in 4 to 5 hours, using Block I satellites.

In the 'kinematic' (in motion) mode, observation time per point is reduced to 45 seconds, but the accuracy attained will not be the same as in the static mode. With kinematics, after initial baseline points are established, one unit stays static (fixed) while other units rove from point to point, without loosing lock on at least 4 satellites. Twenty-five points an hour, using one roving receiver, is not uncommon with kinematic surveying.

Careful planning and scheduling is important for the most cost-effective and efficient GPS surveying. Pre-planning is necessary to overcome site obstructions, avoid 'cycle slips' (gaps in data caused by interruption of the signal between antenna and satellites), and observe as many points as possible during the time period the satellites are visible. Manufacturer software provides the necessary information for GPS survey mission planning. (Printed with permission from Trimble Navigation.)

## Review Questions

1. How many feet are there in an English rod?
2. How many feet are in a Gunters chain?
3. What are the common lengths of steel tapes?
4. In what year was the EDM developed?
5. In what year was the "total station" introduced?
6. Can a range pole or rod conduct electricity?
7. Why should everyone on the crew have a first aid card?
8. Why would you use a dumpy level on a construction site instead of an automatic level?
9. Name three parts of a transit.
10. The bubble moves in the direction of which thumb?
11. Why should the tangent screw be turned into the spring?
12. Name the two types of theodolites.
13. What is a total station?

# Measuring Horizontal Distance

3

## Objectives

After completing this chapter and the field practice the student should be able to:

1. Pace a given distance with an accuracy of plus or minus 1 foot.
2. Demonstrate the proper method of winding and unwinding a steel tape on a reel.
3. Measure between two known points with a steel tape, within third order accuracy.
4. With a steel tape, set a point from a known point at a given distance, within third order accuracy.
5. Correct the distances in objectives 3 and 4 for temperature of the tape and give the corrected distance.
6. Demonstrate the proper handling and care of the retroprism.
7. Demonstrate the use of the retroprism by plumbing over a point:
   A. by hand-held plumbing and
   B. by setting up the prism on a tripod over the point.
8. Discuss critically the advantages of GPS for measuring over long and short distances.
9. Relate the steps necessary to take a position fix using a GPS receiver:
   A. static position;
   B. kinematic position.

---

The first method of measuring distance you should become familiar with is the "pace." Pacing is used to roughly locate points you are searching for. It is also used in taping to prevent you from going too far when moving ahead. To find what your pace is, first set out a measured distance of an even hundred feet (100 ft., 200 ft., etc.). Now pace back and forth along the measured course. Notice the word is "pace" and not "walk" or "saunter." Go at it as if you mean it, in a professional manner. Count the number of paces needed to cover the distance. It is helpful to start with the left foot and count a pace each time the right foot hits the ground. Divide the distance by the number of paces and that is your pace.

## CHAINING/TAPING

The next method to measure distance is taping (also referred to as chaining in many areas). You probably will not use a chain but will use a steel tape instead. Either way, taping is the correct term. Steel tapes come in various lengths and widths, the most common being the 100 ft. and 300 ft. or 30 m. length, in 1/4- and 5/16-in. widths.

Today, most tapes are kept on reels. Years ago, a tapeman was judged on his or her ability to "throw" a tape; not throw it to see how far it would go, but rather, to gather it up in a "figure eight" and by twisting the hands in opposite directions cause the tape to form a loop.

Note that when wound on the reel, the rear loop is put into the slot on the reel with the numbers up and the reel handle on the right (see Figure 3-1). A leather grip or "wang" will always be at

**Figure 3-1** Putting steel tape on a reel

**Figure 3-2** Tape repair kit

the head end to assist your grasp of the tape. The word **chain** has been used many years by surveyors to mean STOP. As the tape is unwound from the reel and it nears the end of the tape, the person with the reel calls out "chain," in a loud voice. When the person at the head end hears "chain," that person must stop immediately. This can also be used as a warning when you get used to it, and every experienced surveyor recognizes that "chain" means "STOP NOW!"

After the tape is unwound, it can be taken from the reel and dragged along the ground as you move forward. Be careful not to let a loop form in the tape. If a loop is pulled tight, the tape will snap and must be repaired with a soldered splice or a patch riveted in place with a tape-repair kit (see Figure 3-2). If you see a loop forming, call out, "chain" to stop the head tapeperson immediately. Work the loop out gently and then proceed. Note that a tape that has been repaired is never as reliable as one that is whole.

As the end of the tape approaches the rear tapeperson, that person should call out "chain." The head tapeperson will stop and the distance can be measured or the point set. The tape should be cleared of any interfering objects between the ends and straightened by flipping the ends.

Some measurements are made on flat ground so the tape can be held flat on the surface. A tension of 10 or 12 lbs. is usually suffi-

**Figure 3-3** Rear tapeperson holding tape flat

**Figure 3-4** Head tapeperson holding tape flat

cient when the tape is fully supported. The rear tapeperson holds an even foot on the point while the head tapeperson clears the tape and checks to see if the point falls within the add foot (see Figure 3-3). If it does not, they call out, "take a foot" or "give me a foot," until the point falls between the zero and the one foot mark. The tape is then stretched and the distance is read. The rear tapeperson calls out the foot and the head tapeperson calls out the tenths and hundredths (see Figure 3-4).

This process is repeated until the last point is reached. The distances are then added up and compared by the head and rear tapepeople. The total is reported to the party chief for recording in the field book. The distances kept by the tapepeople, called **peg notes**, are not allowed by some governmental agencies. In that case, the recorder will record all distances in the field book as they are measured.

On sloping ground or when there are objects that interfere with the tape, it is necessary to raise the tape off the ground and use a plumb bob to project the distances to the point on the ground from the tape. A hand level is used to obtain a **level line** between the head and rear tapepeople, and the tape is then stretched along the line obtained (see Figure 3-5).

It is easiest to sight on the bob string and then hold that point with your fingers. Raise the tape to that point and then lower the bob down to the point on the ground. If you set two points on level ground exactly 100 feet apart with the tape lying flat and then bob up until the tape is supported by the ends only using the same 10 pounds of tension, you will notice that the distance is closer to 100.08 feet. This is caused by the sag of the tape and must be corrected for by pulling harder. You will find that it usually takes about 20 pounds of pull on a 100-foot tape to correct for sag when pulling its full length. Be sure to compare your tape with a standard tape to see what the pull is for your tape and then record the distance in the field book.

To pull a distance with the tape suspended, use the correct tension to raise the tape off the ground to the level line (see Figure 3-6). When the correct foot has been found, the rear tapeperson braces the hand holding the tape with his or her body to allow a solid pull against the force of the head tapeperson's pull. Be careful that the bob string hangs straight down from the tape and is not deflected by the fingers (see Figure 3-7).

When the bob point is directly over the ground point and the string is not swinging, the rear tapeperson calls out, "good" in a loud voice. The head tapeperson must also brace the hand holding the tape to prevent movement. Check the tension of the tape and

## ON SLOPING GROUND

**Figure 3-5** Breaking chain on a slope

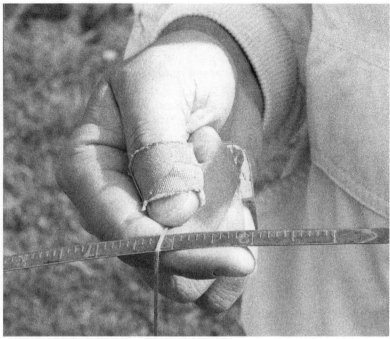

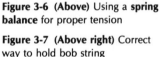

**Figure 3-6 (Above)** Using a **spring balance** for proper tension

**Figure 3-7 (Above right)** Correct way to hold bob string

slide the bob string along the add foot until the point of the plumb bob is directly over the ground point. Make sure that the string is hanging straight and then pinch the string between the thumb and the tape with the forefinger supporting the tape. Slowly let off the tension and read the tenths and hundredths. The rear tapeperson calls out the foot and the head tapeperson calls out the tenths and hundredths. The recorder repeats the numbers, and the process is repeated for the next point.

## TEMPERATURE CORRECTION

Steel tapes are standardized at 68°F or 20°C for foot tapes. For every 15°F over or under that, the 100-ft. tape expands or contracts approximately one hundredth of a foot. The rule of thumb is: One hundredth per hundred per 15°F. A more exact figure can be attained by using the coefficient of expansion of steel which is .00000645 per degree Fahrenheit or 0.0000116 per degree Celsius. The difference in temperature from 68°F times the coefficient of expansion times the distance equals the **temperature correction** to be applied to the taped distance. An example is: $15°F \times .00000645 \times 100.00 = 0.009675$ or approximately .01 feet.

If the tape is cold and you are measuring between two points that already exist, you will read longer than the true distance (see Figure 3-8). If the tape is hot and you are measuring between two points that are already there, you will read short of the true dis-

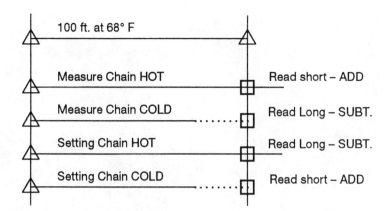

**Figure 3-8** Temperature corrections for chain

tance. When measuring between two points that are already set, the rule of thumb is: Tape cold read long, subtract the correction; Tape hot read short, add the correction. When measuring from one existing point to set another point: Tape cold set short, add the correction, Tape hot set long, subtract the correction. If there is an error in the tape length caused by stretching or shortening due to repairs, it should be treated the same as if the error were due to temperature correction. Sometimes the corrections are in different directions and offset each other.

## BREAKING TAPE/CHAIN

In surveying, as in life, it's not always level ground with the wind at your back. If you cannot get a full tape-length level, it will be necessary to pull the distance in several smaller distances. This is called **breaking tape/chain** and does not mean to see how many short tapes you can make out of a long one; but rather, to break the distances into several smaller distances that can be taped level. The procedure is the same as bobbing up on level ground with the difference that one of the tapepeople will be standing on lower ground than the other (see Figure 3-9).

It is better if the rear or foot end of the tape is on the high end. This allows it to be anchored firmly on the point while the distance is measured on the head end of the tape, which is held level with the rear point on the ground.

## SLOPE TAPING CHAINING

Distance can be measured directly from the instrument head along the slope to allow for longer distances or on slopes that are too steep to work on. This is called **slope taping/chaining** (see Figure 3-10). It requires a person to operate the instrument and read the vertical angle, either as the measurement is being made or immediately before or after. The rear tapeperson holds the foot mark at

**Figure 3-9 (Above)** Rear chainperson breaking chain

**Figure 3-10 (Above center)** Head tapeperson on a break tape

**Figure 3-11 (Above right)** Head tapeperson bobbing over a point on a slope

the point that marks the center of the horizontal axis of the instrument (the vertical motion pivot) while the head tapeperson bobs over the point to be measured (see Figure 3-11).

It is best to pull one distance about knee high and a second distance about shoulder high. After the distances are reduced to horizontal, they are compared and if more than one or two hundredths difference is found, the distances should be remeasured. The vertical angle is read to the point where the head tapeperson is holding the bob string against the tape at the moment of measurement. If the instrument is a transit, the vertical angle is measured from the horizon and can be used as read. If the instrument being used is a theodolite, the angle will be from the **zenith**, or directly overhead, and should be subtracted from 90 or horizontal. The cosine of the vertical angle times the measured slope distance equals the horizontal distance between the two points.

If the tape is passed over the left shoulder, diagonally across the back, and held by the right hand on the right hip while the left hand holds the foot mark against the mark on the instrument, the strength of the whole body can be used to resist the pull of the head tape (see Figure 3-12). Never use only your arm, no matter how strong you are, for leverage is against you. If the pull is still too much to hold against, the instrument operator can hook an arm through your left elbow crook and help to resist the pull.

**Figure 3-12** Rear tapeperson bracing against tape pull

The rear tapeperson must be extremely careful not to bump the instrument and disturb the setup when working near it. When taping from under the tripod legs, be careful not to bump the legs. Brace yourself so as not to be pulled into the tripod by an over-zealous head tapeperson.

**THE EDM**

In today's fast-paced world, the EDM can be used for shorter distances if high accuracy is not required. Most EDMs in use today have a built-in error of $\pm 5$ mm. $+ 5$ parts per million. This translates to an area of uncertainty about the size of a dime, which is usually too much for tight control surveys with short distances. This is not to say that all your distances are off ¼ in. but the possibility exists. Over larger distances, this is distributed and is not as much of a problem. It is difficult to beat the EDM for speed and accuracy on distances of over 200 feet.

**USING A PRISM**

Care must be taken, when using a **prism**, to be certain that the correct prism offset has been set in the EDM. Most prisms in use today have an offset of either 0 or $-30$ mm. Some prism holders allow the prism to be set to one or the other, depending on the setting of the EDM, or for slope distances on slopes over $30°$. Some

**Figure 3-13** Peanut mirror on prism pole

prisms may have an offset of as much as −70 mm.; the offset should be engraved on the back of the prism case as a decimal, so be sure to check any prisms you are not familiar with before use.

### Check the Retroprism Offset

Set three points in a straight line (similar to a peg test of a level) on flat but not necessarily level surface. The distances should be in multiples of the wavelength of your EDM. The manual that comes with the EDM will give the length of the carrier wave. Set the middle point at about two thirds of the distance, not at the center. Set up the EDM over the endpoint and the prism over the middle point and measure the distance. Move the prism to the other endpoint and measure the distance. Move the EDM to the middle point and measure the distance to the prism on the end-point where it is set up. The prism offset will be equal to the over-all distance minus the sum of the two shorter distances.

$$\text{Offset} = AC - (AB + BC).$$

As with any precision tool, it is necessary to take care of the retroprism. The face should be cleaned with glass cleaner and wiped with a soft cloth. Care must be taken not to drop or hit the prism against hard objects.

### Peanut Mirror

The **peanut mirror/1-in. prism** should always be transported in its own leather belt case and not in your carpenters pouch. The larger 2.5-in. mirror (or prism) should be carried in a case or soft pouch designed for it.

Some peanut prisms are mounted on bases that can be screwed to the top of an extendible pole that has a circular fish-eye bubble built in for plumbing the pole (see Figure 3-13). The extension allows the prism to be set to the height of the instrument (HI) to simplify vertical corrections. When plumbing the pole over a point, brace yourself as you would when giving a sight to prevent the mirror weaving around with each breath you take.

Some peanut prism bases have a hook to allow a plumb bob to be suspended below the mirror. Again, brace your hand against your body or use a long lath or range pole section to steady the plumb bob. If possible, the prism holder should be placed directly on the point to reduce movement. Be certain that the mount on a tripod-mounted prism is level and plumbed over the point and that the mirror is pointed directly at the instrument. Some mirror, or prism, mounts are made to be screwed directly to the tripod head. More often, the prism will be mounted on the tribrach to allow the mirror to be replaced by the theodolite or total station with no setup error from the prism.

See Chapter 2 for setup using a tribrach with an optical plummet. As much care must be taken setting up the prism as the instrument. One point is just as important as another. Check to make sure that the line of sight is clear of obstructions. Signal to the instrument operator that the setup is complete and the measurement can be made. Do not move the prism until the measurement has been made and the instrument operator signals you to move to the next point or to point the prism in another direction. If the instrument is to be moved to the tribrach occupied by the prism, turn the swivel knob to release the prism mount from the tribrach, place the prism in its carrying case, and move to the next point. Set the prism mount into the next tribrach and lock the swivel knob, rotate the prism to the point at the instrument position, and proceed as before. Be sure to check that the optical plummet is still centered and the tribrach is level. If it has moved, be sure to tell the party chief about it before the new measurement is made.

The most common type of EDM in use today is the light wave. It uses either an infrared or laser light source to bounce a beam of light off the prism and back to the instrument. There, the phase delay between the transmitted and reflected waves is compared and the distance displayed on the LED screen as feet or meters.

The speed of light (186,411 miles per sec.) is affected by atmospheric conditions. An atmospheric correction must be set into the instrument each day to compensate for changing conditions. Usually, only the temperature and barometric pressure are needed and these can be obtained from the radio weather report. For critical measurements, a thermometer and barometer (set to the correct pressure before starting) are used and new readings are taken before each measurement. The weather report is usually good enough for day-to-day work. Use the card furnished with the instrument to calculate the corrections and set these values into the instrument.

Some EDMs have a built-in target acquisition signal to help locate the prism, but usually the prism will be sighted through the telescope and the hairs centered on the prism. A slight adjustment of the vertical and horizontal tangent screws is used to peak the signal-strength meter. The distance is then shown on the display. Allow the EDM to make several distance corrections before recording the distance to allow for movement of the prism or changes in the light beam's path to average out. If the line of sight is not level, a vertical angle must be read and recorded. Some total stations have automatic slope reduction with the push of a button. It is a good idea to record the distance in both feet and meters in the field book. The two readings can be compared to help locate errors in recording such as transposition of numbers. Some total stations read the information directly into a data collector for interfacing into the main computer back at the office. This should help cut down on human error, but care must be taken to protect the data collector or the days work could be lost. Remember Murphy's Law, "What can go wrong, will!"

## KINEMATIC GPS

**Kinematic GPS surveying** is done using a mobile receiver that has been calibrated to a known position.

### HIGH PRODUCTION *KINEMATIC* GPS SURVEYING

*by Ken Mooyman and Cheryl A. Quirion*

#### Kinematic Field Procedures

To be able to post-process "stop and go" kinematic survey data, lock must be maintained on a minimum of four satellites between all measurement epochs. However, it is not required that lock be maintained on the same four satellites throughout the survey. When doing a kinematic survey the receiver must be able to tell you if satellite tracking has dropped below four. In the case of Trimble receivers, an alarm is sounded to inform the field person that less than four satellites are being tracked; the receiver then requests that it be returned to the last observed point in the traverse. Returning to a known station is necessary in order to reestablish the bias terms in the post processing phase. If the survey is continued without returning to a known station, you will not be able to process the data unless you occupy some other known point later in the survey.

#### Kinematic versus Static

The bottom line of GPS surveying is how to meet the required accuracy at the lowest cost. Using kinematic techniques, it is possible to stay on site for $\frac{1}{60}$ of the time that was necessary for static surveys. However, network geometry and quality control specifications may require that each station be visited more than once in a kinematic survey. For many applications, the kinematic station should have a direct measurement to three reference stations. Restrictions due to logistics must also be considered. Kinematic GPS offers a great time savings over static methods even when these restrictions are considered. On jobs where second-order results were required, kinematic GPS has yielded savings of one to five times compared to static GPS methods.

Even if kinematic methods can only be used on 20 percent of the survey stations, the overall cost per point will be significantly lower. For example, some Trimble users have already incorporated kinematic methods when surveying down power lines, pipeline rights-of-way, and freeway systems. A combination of kinematic and static GPS has become the norm for many GPS users today.

### Pre-Mission Planning

As was stated previously, the minimum number of satellites required to continue a kinematic survey is four. In practice, five or more satellites should be used for better accuracy in post-processing and more flexibility in the field. For instance, if when observing five satellites, a cycle slip occurred on one satellite, you would not have to return to a known point; the survey could continue. In California, there is approximately four hours of five-satellite surveying available. With the restriction of maintaining lock on four satellites, the short working window, and the fast rate of production, pre-mission planning is the key to yielding a profitable kinematic survey. In static GPS surveying a usable station is one with a relatively clear view to the satellites. Kinematic GPS surveying has the same requirements for the survey points; however, it has the additional requirement that the routes between stations must be relatively clear of obstructions. The limits of the survey area must be well known. Reconnaissance is mandatory; the shortest route between two points may not be the best one. If there are certain areas that you cannot drive to while maintaining lock on four satellites, a static survey may be required in those obstructed areas. On kinematic surveys over large areas, it is advantageous to set up a static network of points surrounding the job site. This strengthens network geometry, allows for checks between kinematic and static measurements, and gives you the flexibility in the field to return to one of the static stations in a loss of lock situation. Reducing the chances of cycle slips in the field should be foremost on the kinematic surveyor's mind.

### Quality Control

"Stop and go" kinematic surveys produce coordinate differences in position between the reference station(s) and roving receiver(s). With proper procedures, relative accuracies between the reference station(s) and the roving receiver(s) of 1 centimeter ($\pm$ 2

GPS reference station. *Courtesy of Trimble Navigation*

GPS roving receiver. *Courtesy of Trimble Navigation*

ppm) can be obtained from the existing equipment when observing five satellites for a one-minute observation period. On a line-to-line comparison, the results of a kinematic baseline are virtually identical to its static counterpart for baselines less than 25 kilometers.

. . . As with static GPS surveying, the network geometry is the basis of quality control. Setting up a good network allows for checks; loop closures, repeatability, redundancy and blunder detection. Field procedures should be carried out not only with logistics and production in mind but also field checks; occupying existing static stations, having different roving receivers occupy the same station, and comparing kinematic results with traditional survey techniques.

### Applications for Kinematic GPS

. . . Possible future applications of kinematic GPS include aero-triangulation, survey layout, high-accuracy vehicle tracking, automated profiling of roads and railways, and laser profiling. The aero-triangulation solution would entail correlating GPS timing and position with the photogrammetric camera's focal point at the instant of exposure. This allows photogrammetric mapping to be done with little or no ground control since the position and orientation of the camera is known. Surveying layout requires real-time data links between the fixed and roving receivers. This application will allow for the location of monuments at predetermined positions. Highly accurate vehicle tracking also requires real-time data links between the fixed and roving receivers. Tracking can be used to monitor movement of vehicles.

Automated profiling yields the same product as the vehicle tracking application (i.e., the path of the vehicle's travel) but in the case of profiling the interest lies in the vertical dimension more than the horizontal. This application does not, however, require real-time radio links as does the vehicle tracking problem. It does allow one to essentially digitize the path of travel while driving although the position information is acquired via post-processing techniques. Another application is the possibility of interfacing GPS with other positioning systems such as laser profiling in order to provide accurate ground profiles using helicopters or other aircraft.

*Used by permission of Trimble Navigation, L.T.D.*

GPS vehicle tracking. *Courtesy of Trimble Navigation*

## TO OFFSET AROUND
## AN OBJECT

When using a total station or EDM, an obstruction on the survey line must be overcome. When measuring distances along a line, sooner or later it will be necessary to offset around an object that cannot be moved or cut down. When you are taping a distance and high accuracy is not necessary, the tape can be slightly deflected and the measurement of the line continued if the instrument operator can still give line. If an accurate distance is needed, however, or if the line of sight is obstructed, the line must be routed around the obstacle.

First, the instrument operator should try to find a natural object on the line in the distance that can be seen from the other side of the obstacle. Try to pick an object exactly on the line that has a distinct mark and does not appear to change position as you move around. This "natural sight" will allow the instrument operator to check for a deflection in the alignment after the instrument has been moved around the obstructing object. A telephone pole or a unique tree make good "naturals." A truck or something else that can move or change position should not be used. Also, if using a cylindrical sight such as a power pole or tree, be careful of changing lighting conditions, known as "phase error," that cause the line between light and dark to move around the pole as the light moves. If you can find more than one natural sight on line, so much the better.

Next, the head tapeperson, or the person with the prism, should try to pick a point that can be seen from both sides of the obstacle. Try to pick a point that will give foresights and backsights of equal length. An isosceles triangle is the easiest shape to use and requires fewer calculations of angles and distances (see Figure 3-14).

When turning off (measuring) the angle to offset around the obstacle, again try to find as long a sight as possible. If you cannot find a "natural," have the tapeperson set sights ahead and back along the offset line as far as possible. Unnecessary short sights should be avoided whenever possible. Once the offset point is set and the angle and distance measured, the instrument is moved ahead to the point, and the process repeated to get back on line.

The **deflection angle** at the offset point is double the deflection

**Figure 3-14** Angle offset around obstacle on line

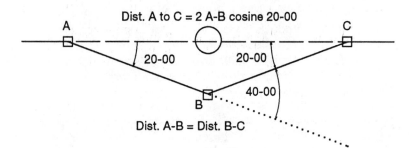

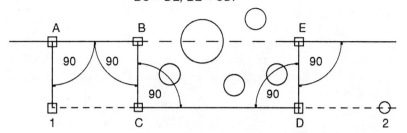

**Figure 3-15** Right angle offset around obstacle

angle turned off the original line. Measure the same distance back to the point on line ahead of the obstacle and move the instrument ahead to the line point. The same deflection angle is now turned that was turned at first point but in the opposite direction. Check to see if the projected new line hits the natural sight that was picked before the offset was made and proceed as before the obstacle was encountered.

If it becomes obvious that there are enough obstructions on the true line to make measurement too time consuming or inaccurate, it may be necessary to establish an offset line parallel to the line. The same precautions should be used as to length of foresights and backsights. It is sometimes possible to set several offset points along the back line. These points can then be sighted on to check for parallelism of the two lines (see Figure 3-15).

**MEASURE BY GPS**

The ideal method to measure horizontal distances between points that are not intervisible or when extreme accuracy is required is now available. If an unobstructed view of the sky is available, the distance can be measured by GPS (see Figure 3-16) stations that make use of signals transmitted from 18 NAVSTAR satellites that orbit the earth.

One GPS receiver is placed on the known control point to establish the relationship of an unknown point to a known point (see Figure 3-15). Another receiver is placed on the point whose position is to be established. Data is collected at both points simultaneously from the satellites for about one hour. This establishes a vector base line between the two points. The latitude, longitude, and elevation of the unknown point can be established to a great degree of accuracy from this information. At least two GPS receivers, a portable computer, and a printer are necessary to locate points by GPS. More receivers help to speed up the work and allow more points to be established at the same time.

**Figure 3-16** Global positioning system. *Courtesy of Trimble Navigation*

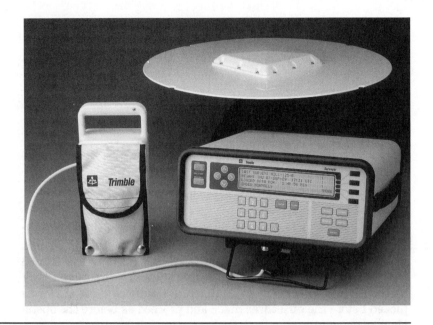

## SITE RECONNAISSANCE

The site reconnaissance is the first step in measuring a distance by GPS. Four pieces of information are needed:

1. the approximate position of the point to be located;
2. the scale;
3. the approximate latitude, longitude, and elevation off a good map of the area (such as a United States Coast and Geodetic Survey topographic quad sheet); and
4. select a point that will give good satellite visibility.

An elevation view of 15° above the horizon is recommended (see Figure 3-17). Make a plot of obstructions around the point. Show **azimuths** or bearings from the point and approximate height of each obstruction. Make notes on the date the information was taken, a good description of the site, the amount of time needed to travel to the site, how to locate the site, and information on locked gates or people that can be contacted for permission to enter or open gates. If you find a gate closed, be sure to close it after you have gone through. It helps to keep the landowners happy.

The receiver will give the best time to try to take the measurements on a printout of known satellite positions. Try to pick a time when the most satellites are visible from all locations.

Be sure the receivers have a clear view of the sky at all locations. The reference position is entered into the receiver at the known point.

The correct satellite positions are broadcast by the satellites every fifteen minutes. This information is received directly into

**Figure 3-17** Unobstructed view of the sky

your receiver. Be sure to allow time to update your information before making the measurement. Set up your receiver according to the manufacturer's instructions and get the satellite schedule plot so that you can schedule the best time to make the measurement. All stations must make observations at the same time on the same satellites.

Each operator needs a planning worksheet containing all the information needed to make the operation successful. Coordination of observations is essential. Allow time for travel and setup as well as the time to make the observation and break down the equipment to move to the next position. Make sure you have everything you need before leaving the office. If you forget something, the whole project may need to be scrubbed and rescheduled. Allow yourself plenty of time to get to the work site. Remember Murphy!

Set up the antenna over the point and connect the cord to the receiver and the antenna (see Figure 3-18). Check the setup and

**Figure 3-18** GPS remote station. *Courtesy of Trimble Navigation*

**Figure 3-19** GPS receiver. *Courtesy of Trimble Navigation*

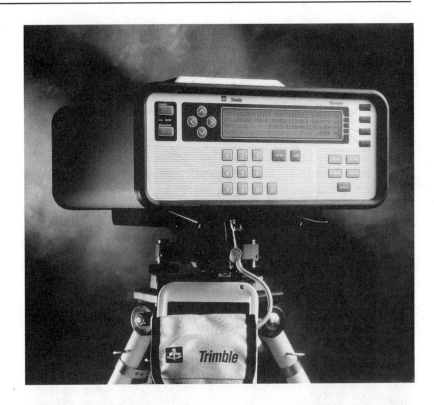

connections to make sure everything functions properly. Be sure to measure the antenna height above the point and record the measurement. Check the time and set up your computer or data collector. Start the receiver and wait until the time to shut down the observation before moving anything. After the observation, take down the antenna (after again checking the height above the point) and proceed to the next point or back to the office to have the information processed.

If the information has been recorded on computer disks, be sure to protect them from damage or all the information may be lost. Magnets and fingerprints are unhealthy for computer disks. Treat them as well as you treat your favorite music CDs.

The computer will print out the latitude and longitude for the new point as well as the elevation (see Figure 3-19). This information can be converted into state plane coordinates from which the distance and bearings between the points can be calculated to a great degree of accuracy.

We've come a long way from ropes to satellites but we still do the same thing: measure distances from point to point better than anyone else!

# Review Questions

1. When you place a tape on the reel, which way should the numbers face? Which side should the reel handle be on?

2. What should you do when you hear someone call "chain"?

3. What is the normal pull for a 100-ft. highway tape?

4. What is the standard temperature for a steel tape?

5. What is the correction for 1° difference in temperature from standard?

6. What does the term "breaking tape" mean?

7. When breaking tape, how do you determine if the two ends are level?

8. What trigonometric function is used times the slope distance to get the horizontal distance when "slope taping"?

9. What does EDM mean?

10. Why do some surveyors use a tape for distances under 200 feet?

11. Why is it necessary to check the prism offset on a new prism before using it in the field?

12. When setting a prism over a point using a tripod and tribrach, how should you proceed?

13. When offsetting around an obstacle, what should you do before setting the offset point?

14. What do you check for when setting a GPS station point?

15. If you must go through a closed gate on the way to the station point, what should you do?

16. You need the distance between two found property corners. Using a 100-ft. steel tape you measure 100.54 and 97.16. The temperature is 87°F. The tape is 100.000 ft. at 20 lbs., which you used in taping. What is the actual distance between the corners?

17. The plans call for a point to be set exactly half way between the property corners in question 16. What distance should should you pull to set a hub on line between the corners?

18. A distance of 593.076 is measured with an EDM at a 72°27'15" vertical angle. What is the horizontal distance?

19. You need to set a point at 426.30 ft. from the instrument. The first trial point is set at a vertical angle of 94°26'20" with a distance of 424.15 feet. How much farther should you set the point?

20. While running a line through timber, you have found a large tree on line. The tapeperson sets a hub on an offset point. The deflection angle to the offset is 15°26'13" and the distance is 215.20 feet. From the offset hub, what angle should you turn to set the new point on line and how far from the offset should it be set?

21. What is the center line distance at the point on line from the last center line point in question 20?

22. List the operations necessary to set up and take a GPS position of a station point.

# Measuring Vertical Distance

**4**

## Objectives

After completing this chapter and the field practice, the student should be able to:

1. Differentiate between a level line and a horizontal line.
2. Demonstrate the proper use of a hand level:
   A. to find a difference in elevation and
   B. to move up a slope.
3. Demonstrate the proper way to set up a level.
4. Demonstrate the proper way to plumb a rod on a bench mark or turning point.
5. Distinguish between a good turning point and a bad turning point.
6. Reduce the notes for a differential level circuit and show closing error.
7. When working as a rod carrier on a level party, complete a level run to third order accuracy.
8. When working as the instrument operator of a level party, complete a level run to third order accuracy.
9. Apply the proper note form to record the field data for a level circuit, reduce the notes, and give the order of accuracy.
10. Use trigonometric methods to find the difference in elevation between two given points.

We normally consider vertical distances as differences in elevation above and below sea level. The term "sea level" is actually "mean sea level," a figure that is arrived at by a series of observations of the rise and fall of the tide taken over a period of 18.6 years. The latest adjustment of this information is known as NAVD 88 (North American Vertical Datum 1988).

**MEAN SEA LEVEL**

## CURVATURE AND REFRACTION

If we measure the vertical distance between two points along a plumb line we refer to the result as the difference in elevation. A **level line** is a line that curves to follow sea level. Do not confuse a level line with a **horizontal line** (see Figure 4-1). A horizontal line forms a 90° angle with a vertical line. A level line is not flat because it follows the curvature of the earth's surface. The difference is not great (only a little over 0.02 ft. in 1000 ft.) but is extremely important if you are called on to lay out a flat surface as opposed to a level surface. Another factor that will affect a level line is **refraction.** As light waves pass through the earth's atmosphere, they are bent downward. As a result, for precise measurements of differences in level, a correction must be applied for **curvature** and refraction. The formula for the combined corrections is:

$(c + r)$ in feet = 0.574 times the distance in miles squared
$(c + r)$ in meters = 0.0675 times the distance in kilometers squared

**Figure 4-1** Horizontal line and level line

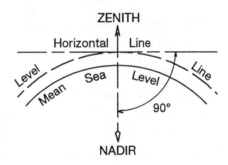

**Figure 4-2** Curvature and refraction

$c+r = 0.0675\ K^2$ (c+r) in meters K in km

$c+r = 0.544\ M^2$ (c+r) in ft. M in miles

$c+r = 0.206\ T^2$ (c+r) in ft. T in Thous. ft.

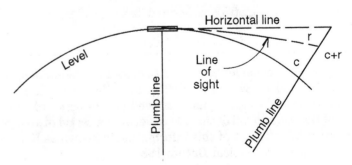

(see Figure 4-2). If shots are kept to 300 ft. or less, the correction can be ignored for most work. Some government agencies require that **turning points** be no more than 400 ft. apart when running levels.

---

Rough differences in elevation can be measured with a hand level. The distance from the ground to the observer's eye should be known. This is known as your **personal HI** and can be used to judge differences in elevation by looking through the hand level at a point on the slope when the bubble is centered on the cross hair. This point is the elevation of the line of sight through the level:

$$BM + BS = HI.$$

Move up to that point and repeat the operation as many times as necessary to reach the point where elevation is needed. This method is useful for extending levels beyond range of the instrument when taking **cross sections**. The hand level can also be used with the level rod or a folding ruler to measure more precise differences in elevation. Hold the hand level next to the rod, adjust it up or down until the cross hairs are on the point, and then read the difference in elevation from the point on the rod next to the hand level (see Figure 4-3).

**THE HAND LEVEL**

**Figure 4-3** Using a hand level with a rod

## THE ALTIMETER

An **altimeter** can also be used for approximate elevation between points (see Figure 4-4). Set the altimeter to the elevation of a known point and then move it to the new point and read the elevation. More precise elevations can be obtained by using two or more al-

**Figure 4-4** Altimeter

timeters. First, set all the altimeters to the elevation of the known **bench mark**. Leave one altimeter there with an operator to read any variations in the needle caused by changes in barometric pressure and record the time of the change. The other altimeters are transported to the remote points and the elevations recorded along with the time they were read. The remote altimeters are then taken back to the starting bench mark and compared with the control altimeter and the readings **prorated** to coincide with the readings of the stationary unit.

---

The most common method of measuring differences in elevations is known as **differential leveling**. In differential leveling, a reading is taken through a level sighted on a graduated rod held on a point of known elevation. The rod is then moved to a point where the elevation is needed and a second reading is taken. The difference in the two readings is the difference in elevation between the two points.

Differential levels always start from a known elevation. Whether this elevation is a bench mark of true elevation above or below sea level or a "temporary" bench mark with an assigned elevation used for this job only, each level circuit begins with an elevation. The level rod is held on the starting bench mark and the level is set up in a position that allows a clear view of the rod face. Try to pick a position that does not require the rod to be read in the top or bottom tenth; allow yourself a little room for setup error.

If you are using a tripod with wing nuts to tighten the legs, be sure to release the wing nuts and retighten them to release tension after the legs are set in the ground. If the tension is not relieved before leveling the instrument, the legs might adjust themselves without your noticing during the level shot and cause an error in the reading. Constant-tension legs such as are used on a theodolite should be adjusted, using the allen wrench provided with the tripod, until each leg slowly settles from its own weight when extended horizontally and released.

Once the tripod is set, the level head is leveled, using the bubble or fish-eye bubble, and is then focused on the rod. When ready to take the reading, the instrument operator raises one arm over his head as a signal to the rodperson to plumb the rod on the point. The rodperson should stand directly behind the rod, facing the instrument, and return the signal. The rod is then placed on the point and rotated slightly to make sure it is sitting freely on the point. The rod is plumbed by either balancing it gently between the fingers or by using a rod level (see Figure 4-5).

If the reading is near the top of the rod, the instrument operator may signal the rodperson to "wave the rod." This is done by raising

## DIFFERENTIAL LEVELING

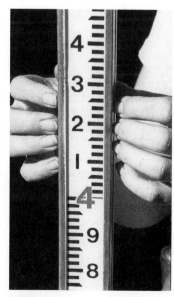

**Figure 4-5** Balancing rod with fingers

**Figure 4-6** Waving the rod

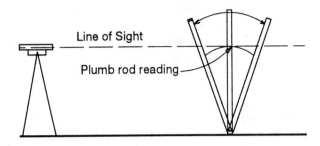

one hand and waving it toward the rodperson (see Appendix A, Surveyor's Hand Signals). The rodperson will then rock the rod slowly back and forth. Because a straight line is the shortest distance between two points, if the lowest reading is taken while the rod is being waved, it will be taken when the rod is exactly plumb (see Figure 4-6). Once the instrument operator is sure that the rod reading is correct, the reading is recorded, and the instrument operator waves both arms overhead as a signal to the rodperson to move ahead to the next point.

The next point may be a "turning point" or just an elevation shot or foresight. If the shot is to be a turning point, care must be taken in its choice. A good turning point is on a specific point that allows only one point to contact the foot plate of the rod (see Figure 4-7). If the rod is placed on a flat surface, it will give a false reading when waved back and forth. Check for a point by rotating the rod; if it spins on the point and does not seem to move or walk from the point, you have picked a suitable turning point.

If a good natural point cannot be found in the area, a **hub** can be driven in at an angle to make a point. In soft soil, the plumb bob can be pushed into the ground and used for a turn; but be sure not to pull out the point too soon. If a **blunder** has been made or if the instrument is picked up at the wrong time, you must go back to the last identifiable turning point. If you are pulling your turns, then you must go all the way back to the bench mark and start again.

**Figure 4-7** Turning point for rod

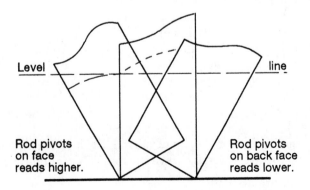

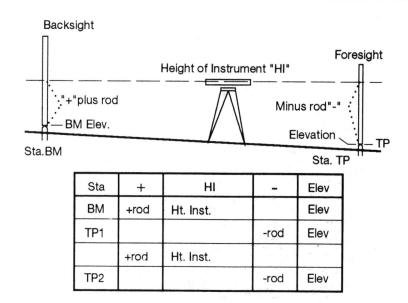

**Figure 4-8** Typical level setup

| Sta | + | HI | − | Elev |
|-----|-----|---------|------|------|
| BM | +rod | Ht. Inst. | | Elev |
| TP1 | | | -rod | Elev |
| | +rod | Ht. Inst. | | |
| TP2 | | | -rod | Elev |

Once the elevation of the turning point has been established by again reading the rod and recording the reading in the field book, the instrument may be picked up and the whole process repeated. To summarize the procedure for one turn (see Figure 4-8):

> elevation plus the backsight reading gives the
> height of the instrument (HI) minus the foresight rod
> reading equals the new elevation.

On each setup, only one **plus rod** or backsight is taken on the known elevation. However, any number of **minus rods** or foresights may be taken. A common error of new level operators is to call any shots read between the level and the bench mark backsights because they are behind the instrument in the direction of travel and any shots taken beyond the level foresights, "because that's the direction we're going" (see Figure 4-9).

A plus rod is taken on the known elevation to establish the height of the level above the datum. This height is known as the **HI**, or height of the instrument, and is a plane that passes through the cross hairs of the level. As the level is rotated, the plane is seen through 360°. When the rod is read anywhere in this plane other than the plus rod, a distance above or below this plane can be read with the level.

Most rods will be below or minus rods that are used to establish elevations on points on the surface of the earth or for construction features. However, some shots will be taken above the plane to establish elevations of ceilings or top of tunnel elevations called **obverts** (see Figure 4-10).

**Figure 4-9** Multiple minus rods

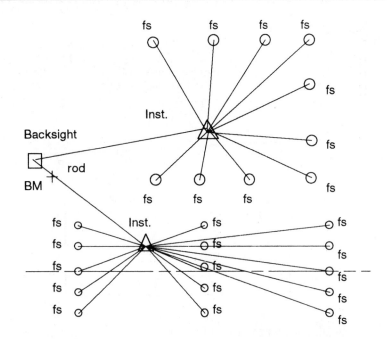

**Figure 4-10** Obvert shot

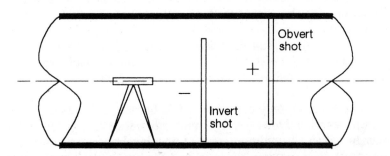

If the rod is below the plane, the reading is subtracted from the HI to give the relative elevation. If the rod extends above the plane, the reading is added to the HI to give the relative elevation. The line of sight through the cross hairs is perpendicular to a plumb line on the earth's center for only a short distance and then begins to bend downward because of the refraction of the atmosphere. Also, the level itself may not be in perfect adjustment.

**Balancing the Shots**

A method used to correct such errors is called **balancing the shots.** Any error from a flat plane will cause the line of sight to form a cone as it is rotated around the axis (see Figure 4-11). If the backsights and foresights are taken at very nearly the same distance from the instrument, this error will be canceled out. Balancing the

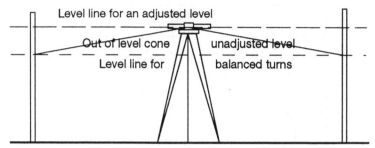

**Figure 4-11** Balanced turns

Balanced turns compensate for out of level

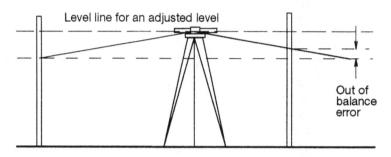

**Figure 4-12** Out of balance back-sight and foresight

shots is accomplished by the rod carrier pacing the distance to the instrument as they move ahead and then pacing ahead approximately the same distance to the turning point. On sloping ground this can be done by traveling diagonally across the slope (see Figure 4-12).

**NOTE KEEPING**

Note keeping for differential leveling has developed into a fairly standard form (see Figure 4-13). If the plus shots and HIs are kept on one line and the minus shots and elevations are kept on the next line, it leaves room to make corrections in the blanks between the lines. When making corrections in the field, it is good practice not to erase the erroneous entry. Instead of erasing, draw a line through the bad entry and write the correct entry above it in the space left by skipping a line. Also regarding this subject, any changes made in the office should be made in red pencil to show that they are not made in the field.

If the levels have been run in a loop back to the starting bench mark, called a **level loop,** the total of the plus rods should be equal to the total of the minus rods. If the levels have been run as a connecting level run, the difference between the plus rods and the minus rods should be equal to the difference in elevation between the starting and ending bench marks.

A level run must never be left hanging open. It must either be closed back to the starting bench mark or closed into another

**Figure 4-13** Differential level notes

| Street Profile | | | | | |
|---|---|---|---|---|---|
| STA. | + | H.I. | − | ELEV. | PROF. |
| B.M. | | | | 438.21 | |
| | 3.71 | 441.92 | | | |
| T.P. | | | 5.62 | 436.30 | |
| | 5.10 | 441.40 | | | |
| 0+00 | | | 5.2 | | 436.20 |
| 0+50 | | | 5.3 | | 36.1 |
| 1+00 | | | 5.5 | | 35.9 |
| 1+50 | | | 5.7 | | 35.7 |
| 2+00 | | | 7.3 | | 34.1 |
| 2+50 | | | 7.4 | | 34.0 |
| 3+00 | | | 5.6 | | 35.8 |
| 3+50 | | | 5.9 | | 35.5 |
| 4+00 | | | 6.1 | | 35.3 |
| 4+50 | | | 6.4 | | 35.0 |
| 5+00 | | | 6.6 | | 34.8 |
| T.P. | | | 6.37 | 435.03 | |
| | 4.31 | 439.34 | | | |
| 5+50 | | | 5.0 | | 434.3 |
| 6+00 | | | 5.2 | | 434.1 |
| 6+37.52 | | | 5.7 | | 433.6 |
| T.B.M. | | | 5.31 | 434.03 | |
| | | | | | |
| | | | | | |

bench mark of known elevation. Any side shots or minus rods that are not turned through cannot be checked for error. This leads to a good rule to follow: always turn through any point that must have an accurate elevation placed on it.

If closure is not 0.00 or "flat" but is within the allowed degree of accuracy (see Figure 4-14), the amount of error can be adjusted out by simple proportion. If you have an error of 0.04 in closure and have turned through three temporary bench marks, each TBM would receive a correction of 0.01.

When levels must be run over long distances, it is sometimes advisable to make short loops and close back to a TBM that you have set along the line. Closing short loops allows you to catch an error in the level run before you have traveled too far. Another method of running a long level run is **double rodding**. This saves the return run and you only have to walk half as far. Start out as usual with a plus rod to get the HI. On the minus rod, find two

**Figure 4-14** Field procedures table of level closures.

| Order<br>Class | First<br>I | First<br>II | Second<br>I | Second<br>II | Third |
|---|---|---|---|---|---|
| Minimal observation method | micrometer | micrometer | micrometer<br>or 3-wire | 3-wire | center wire |
| Section running | SRDS<br>or DR | SRDS<br>or DR | SRDS<br>or DR« | SRDS<br>or DR* | SRDS<br>or DR§ |
| Difference of forward and backward<br>  sight lengths never to exceed | | | | | |
|     per setup (m) | 2 | 5 | 5 | 10 | 10 |
|     per section (m) | 4 | 10 | 10 | 10 | 10 |
| Maximum sight length | 50 | 60 | 60 | 70 | 90 |
| Minimum ground clearance of line of<br>  sight (m) | 0.5 | 0.5 | 0.5 | 0.5 | 0.5 |
| Even number of setups when not using<br>  leveling rods with detailed calibration | yes | yes | yes | yes | – |
| Determine temp. gradient for the<br>  vertical range of line of sight at<br>  each setup | yes | yes | yes | – | – |
| Maximum section misclosure (mm) | $3\sqrt{D}$ | $4\sqrt{D}$ | $6\sqrt{D}$ | $8\sqrt{D}$ | $12\sqrt{D}$ |
| Maximum loop misclosure (mm) | $4\sqrt{E}$ | $4\sqrt{E}$ | $6\sqrt{E}$ | $8\sqrt{E}$ | $12\sqrt{E}$ |
| **Single-run methods** | | | | | |
| Reverse direction of single runs every<br>  half day | yes | yes | yes | – | – |
| **Nonreversible compensator leveling<br>  instruments** | | | | | |
| Off-level/relevel instrument between<br>  observing high and low rod scales | yes | yes | yes | – | – |
| **3-wire method** | | | | | |
| Reading check (difference between<br>  top and bottom intervals) for one<br>  setup not to exceed (tenths of rod<br>  units) | – | – | 2 | 2 | 3 |
| Read rod 1 first in alternate setup<br>  method | – | – | yes | yes | yes |
| **Double scale rods** | | | | | |
| Low-high scale elevation difference<br>  for one setup not to exceed (mm) | | | | | |
|   With variable compensator | 0.40 | 1.00 | 1.00 | 2.00 | 2.00 |
|   Other instrument types: | | | | | |
|     Half-centimeter rods | 0.25 | 0.30 | 0.60 | 0.70 | 1.30 |
|     Full-centimeter rods | 0.30 | 0.30 | 0.60 | 0.70 | 1.30 |

*Source:* Federal Geodetic Control Manual, September 1984
(SRDS – Single-Run, Double Simultaneous procedure)
(DR – Double-Run)
(SP – SPur, less than 25 km, double-run)
D – shortest length of section (one-way) in km
E – perimeter of loop in km
« – Must double-run when using 3-wire method.
\* – May single-run if line length between network control points is less than 25 km.
§ – May single-run if line length between network control points is less than 10 km.

**Figure 4-15** Double rod leveling
notes

| Double Rod Levels | | | | | 4-16 |
|---|---|---|---|---|---|
| STA. | + | H.I. | − | ELEV. | |
| B.M.#1 | | | | 536.31 | |
| | 5.21 | 531.52 | | | |
| TP1 H | | | 4.61 | 526.91 | |
| L | | | 5.22 | 526.30 | |
| H | 4.99 | 531.90 | | | |
| L | 5.60 | 531.90 | | | |
| TP2 H | | | 5.51 | 526.39 | |
| L | | | 6.01 | 525.89 | |
| H | 4.72 | 531.11 | | | |
| L | 5.21 | 531.10 | | | |
| TP3 H | | | 4.91 | 526.20 | |
| L | | | 5.82 | 525.28 | |
| H | 4.31 | 530.51 | | | |
| L | 5.21 | 530.49 | | | |
| TP4 H | | | 3.12 | 527.39 | |
| L | | | 4.23 | 526.26 | |
| H | 12.61 | 540.00 | | | |
| L | 13.72 | 539.98 | | | |
| BM#2 H | | | 4.13 | 535.87 | 35.86 |
| L | | | 4.13 | 535.85 | 35.86 |
| | | | | | |
| | | | | | |
| | | | | | |
| | | | | | |

turns close together: a high turn and a low turn. Make sure that there is a difference between the turning points that is significant or the two shots might get mixed up. Record the shots as high and low or 1 and 2 or whatever—just do not confuse them. One set of shots is used as the forward run and the other set is used as the return run, but you only walk half as far and it only takes half the time (see Figure 4-15).

### The Rod Carrier

When the rod carrier and the person running the level have worked together for a time, they should be able to run levels just about as fast as they can walk. The rod carrier should be looking for the turning point as they approach the proper location. Deciding how far to go will come with practice. Extend the rod and when nearing the trial point, raise the rod to a vertical position and watch the instrument person.

### The Instrument Operator

The instrument operator follows the rod and if the rod is too high or too low when the rod is close to plumb, signals the rod carrier to go farther ahead or come back toward the level. As soon as the rod is read, the level operator signals the rod carrier by waving both arms overhead, records the rod, picks up the level, and moves ahead. The rod carrier marks the turning point with a **kiel** and turns so that the rod face is pointing toward the instrument person as they approach the next setup point. The instrument operator should use a hand level to be sure the rod can be seen from the set-up point before leveling the instrument. Once the legs are set and the level is rough leveled, check again through the instrument to be sure that the rod can be seen and then adjust the level and read the rod.

On setup points from which many foresights are to be taken, the instrument person should track the rod carrier with the instrument so that the rod can be read as soon as it comes to the plumb position. As soon as the rod is read, the rod carrier should be signaled so that they can proceed to the next point without delay. The reading should be recorded in the field book while the rod carrier is moving to the next point, and the instrument should be turned in anticipation of the next reading. With practice, the rod carrier will barely need to stop.

When the rod can no longer be seen from the instrument position because it is too high or too low, the instrument person waves one arm overhead in a circular motion to signal the rod carrier to find a point for a turn. The rod carrier finds an appropriate turning point and, placing the rod on the point with the face away from the instrument, returns the signal. When it has been ascertained that the rod can be seen from the instrument position, the face is turned toward the instrument and the shot is taken. This procedure will prevent both reading the rod before the rod carrier is satisfied with the turn and picking up the level before the turn is set.

Some turns are forced out of balance because of an intervening obstacle such as a shot across a freeway or river. To achieve a balanced turn in these conditions, a procedure known as **reciprocal leveling** is used. The minus shot is taken on the far side of the obstacle, the level moved ahead, and the plus shot taken from the point set. The rod is then taken back to the previous turn and a minus shot is taken. The two long shots are then averaged and the result used for the new elevation (see Figure 4-16).

The Lenker direct reading rod is used on construction projects to obtain elevations directly (see Figure 4-17). The "L-E-VATION"

## RECIPROCAL LEVELING

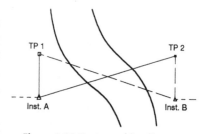

**Figure 4-16** Reciprocal leveling

**Figure 4-17** Lenker rod

rod is a proven short cut in field work. This two-section rod is extendible from 5 to 10 ft. and gives direct readings in elevation. In use, the rod is read on the bench mark and the reading signaled to the rod carrier. A finger tip is placed on the reading and the band released by loosening the locking pin. The band is then moved by sliding with pressure from the thumb and fingers until the elevation of the point is on the point read. The locking pin is then inserted into the nearest hole and locked. The rod should be extended and closed several times to settle the reading and then read again by the instrument person. If the reading is correct, the rod may be moved to the points where elevation readings are needed and read directly for them. If a slight adjustment is needed, the lock is loosened, as the instrument operator watches, and the band is slipped along, until the correct reading is on the cross hair. The instrument person then signals "good" and the rod is locked and checked. Always be sure to check back into a known elevation to be sure that the rod has not slipped after it was set. ALWAYS CLOSE A LEVEL RUN!

## THE PEG TEST

If you have a consistent problem closing the level runs, the level may be out of adjustment. A **peg test for level** is used to test the instrument for being level. To peg test the level, you must find a fairly flat and clear area about 300 ft. long. Using a 300-ft. tape stretched flat for both distance and line, set a stake at the 0-, 150-, and 300-ft. marks. These stakes must be in a straight line with no more than a 4-ft. difference in elevation between the two end stakes. Set the level over the middle stake and level it carefully. Take a rod reading on one end of the stakes (A) and record the reading. Have the rod carrier move to the other end stake (B) and take a rod reading on that stake. Subtract the reading on B from the reading on A. The difference between the two readings is the true difference in elevation between the two end stakes. The balanced shot principle proves that the two end readings are the true difference in elevation between the two end stakes. The balanced shot principle proves that the readings of the two ends will be the same as if the instrument had been in correct adjustment. Now move the level to point A, with the rod still at point B, and set up on the line directly behind A as close as the near focus of the level will permit. Take a rod reading on point B and have the rod carrier move to point A. Take a reading on point A and subtract the rod reading of B from the rod reading on A. The difference between the second difference and the first difference is the amount of correction to be made to the level hairs. Consult the manual for the level you are using for the proper method to adjust the horizontal cross hair to read the true difference in elevation between the two end

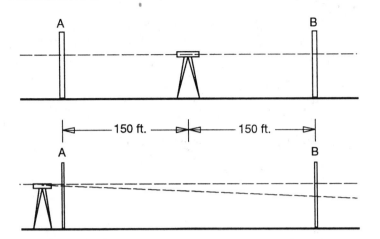

**Figure 4-18** Peg test for level

stakes. It is good practice to repeat the test after adjusting the cross hair to be sure that the instrument is now in proper adjustment (see Figure 4-18).

**TRIGONOMETRIC LEVELING**

The terrain may be too varied or steep to make differential leveling practical. The readings are limited by the length of the rod. Although some rods are made that will extend to 25 ft. or more, eventually, even that is not enough. One method of dealing with extreme differences in elevation is to use **trigonometric leveling.** Do not panic if you have not had trigonometry yet. The formulas to reduce the readings will be supplied here, but it is essential that you learn trigonometry if you want to progress as a surveyor and now is as good a time as any. First, the instrument person must know how to read vertical or zenith angles on a transit or theodolite. Next, the distance between the points must be known or a method of measuring them must be found.

**STADIA**

One method of measurement is the **stadia shot** or reading the distance by using the upper and lower cross hairs in the instrument. This reading is usually in a ratio of 1 : 100. The vertical distance from the HI to the center hair is calculated by multiplying the stadia reading by 100 times the cosine of the vertical angle times the sine of the vertical angle (see Figure 4-19).

Vertical difference $= 100 \times$ stadia intercept $\times \cos \theta \times \sin \theta$

The rod reading must be subtracted from this elevation to find the elevation of the point. If the HI is set on the center cross hair,

**Figure 4-19** Incline stadia shot

Vertical diff.= 100 x stadia x cos$\theta$ x sin$\theta$

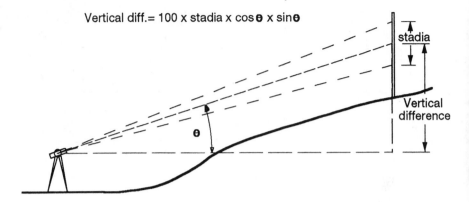

the difference can be added or subtracted directly from the station elevation (see Figure 4-20). The horizontal distance is calculated by multiplying the stadia interval by 100 times the cosine squared of the vertical angle (see Figure 4-21).

$$\text{Horizontal distance} = 100 \times \text{stadia intercept} \times \cos^2\theta$$

The effects of curvature and refraction must be calculated and corrected for on shots over 1,000 feet. Trigonometry levels are only as accurate as the care taken in their use. In 1988, a group of surveyors used a GPS survey to check the elevation of Mt. Rainier in Washington against the official elevation of 14,410 ft. that had been measured using trigonometry levels from several stations. The new reading was only 1.1-ft. higher, proving that work done using careful measurements by trigonometric methods is accurate

**Figure 4-20** Shooting HI

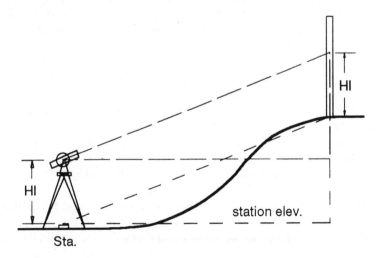

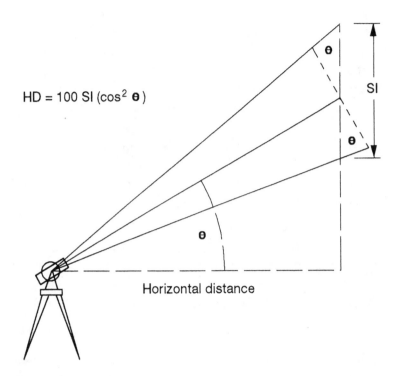

**Figure 4-21** Horizontal distance calculation

$$HD = 100 \: SI \: (\cos^2 \theta)$$

SI

Horizontal distance

enough for most purposes. The difference in elevation derived from using total stations and mirrors to measure the slope distance and vertical angle is more accurate than stadia measurements. The measured distance or slope distance can be multiplied by the sine of the vertical angle to derive the vertical distance. Zenith angles need to be converted into vertical angles. Some instruments have a built-in computer that gives the vertical and horizontal distances directly.

**GLOBAL POSITIONING SYSTEM**

The use of the Global Positioning System (GPS) array of satellites to give position and elevation is the latest method of establishing elevations over extremes of vertical difference. When the master unit is set on a known position and the altitude entered into the unit, the elevations of the remote stations can be calculated to a greater degree of accuracy than was previously possible. The concept of remote sensing is revolutionizing the industry, and by the time this book is available, these methods may already be obsolete. Auto-levels were a new concept to the surveyors of the 1950s. Now they are using satellites. One wonders what methods you will be using thirty years from now. You should appreciate the age you live in and make the most of it.

## Review Questions

1. How is mean sea level determined?
2. What is the latest adjustment of the vertical datum?
3. How far apart can turning points be to avoid curvature and refraction correction?
4. What is your personal HI?
5. How would you use a hand level to measure a difference in elevation between the top and bottom of a slope?
6. How many altimeters are needed to get the elevation of a remote point or points?
7. What are the five steps in differential leveling between two points?
8. What makes a desirable turning point?
9. How many back sights can be taken from one setup?
10. How many fore sights can be taken from one setup?
11. Why should turns be balanced?
12. Why is it good practice to leave a line in the field book between the plus and the minus rods?
13. Should you erase an erroneous entry?
14. How would you check for a math mistake in reducing the notes for a level loop that closes back to the same BM?
15. How would you check for a mistake in a connecting level run?
16. Why should all important points be turned through?
17. What do you call the procedure for leveling across a river that prevents balancing the turning points?
18. What is a direct reading rod called?
19. What is trigonometric leveling?
20. How does GPS give the elevation of a point?
21. Reduce the following field notes and adjust out any error of closure.

| STA. | + | HI | − | ELEV. |
|------|------|------|------|------|
| BM# 1 | | | | 80.82 |
| | 5.42 | | | |
| T.P. 1 | | | 7.62 | |
| | 2.92 | | | |
| T.P. 2 | | | 10.28 | |
| | 1.03 | | | |
| T.P. 3 | | | 11.61 | |
| | 0.76 | | | |

| STA. | + | HI | − | ELEV. |
|---|---|---|---|---|
| T.P. 4 | | | 12.02 | |
| | 0.42 | | | |
| T.P. 5 | | | 9.70 | |
| | 0.27 | | | |
| TBM# A | | | 7.36 | |
| | 10.26 | | | |
| T.P. 6 | | | 1.20 | |
| | 11.37 | | | |
| T.P. 7 | | | 0.37 | |
| | 12.22 | | | |
| T.P. 8 | | | 0.74 | |
| | 9.60 | | | |
| T.P. 9 | | | 1.84 | |
| | 11.67 | | | |
| BM# 1 | | | 3.22 | |

What is the adjusted elevation of TBM# A?

22. Reduce the following field notes.

| STA. | + | HI | − | ELEV. |
|---|---|---|---|---|
| BM# 101 | | | | 1246.37 |
| | 4.26 | | | |
| T.P. | | | 7.30 | |
| | 6.32 | | | |
| T.P. | | | 4.92 | |
| | 7.98 | | | |
| T.P. | | | 8.73 | |
| | 3.33 | | | |
| T.P. | | | 10.34 | |
| | 8.64 | | | |
| T.P. | | | 5.47 | |
| | 4.18 | | | |
| T.P. | | | 6.02 | |
| | 10.08 | | | |
| BM# 102 | | | 1.13 | |
| BM# 102 | | | | True Elevation = 1247.24 |

What is the closing error?

23. The EDM is set up over a hub with a known elevation of 542.20. The measure up (MU) to the instrument is 5.26. A zenith angle of 72°20′30″ is turned to a prism that is mounted on a tripod 4.72 above a new hub. The slope distance is measured as 576.23. What is the elevation of the new hub and how far is it horizontally from the original hub?

24. Using a 300-ft. tape, three hubs, a level, and a rod, check the level for adjustment using the peg test. Do not adjust the level without the permission of the instructor.

25. Using a level and a rod, make a level run of at least a half mile in length. Reduce the field notes and check for closure. Use proper field and office procedures and turn in your completed notes.

# Horizontal and Vertical Angles

**5**

## Objectives

After completing this chapter and the field practice, the student should be able to:

1. Demonstrate finding a point perpendicular to a given line using the wind angle or wing-ding method.
2. Demonstrate finding a point perpendicular to a given line using a right angle prism.
3. Construct a right triangle using the 3-4-5 principle.
4. Given the lengths of two walls of a building, calculate the diagonals to check for squareness.
5. Given a transit or a theodolite and five points around the horizon, close the horizon within plus or minus 30 sec. of arc.
6. Given a transit or a theodolite and two points of different elevation, find the vertical angle between the two points.
7. Demonstrate the ability to set up a transit and/or theodolite on uneven ground within industry-acceptable time limits.

---

In the previous chapters we have learned how to measure distances in a horizontal and vertical direction. Now we will learn how to measure a change in direction, known as an angle, between two lines.

The simplest method of turning an approximate angle is the **wind angle** or **wing ding**. Stand on the point that you wish to set a perpendicular to and face so that your extended arms point along the forward and back tangents of the line. Stand erect with your

**THE WIND ANGLE OR WING DING**

**Figure 5-1** Wing ding. *Courtesy of Michael Perniciaro*

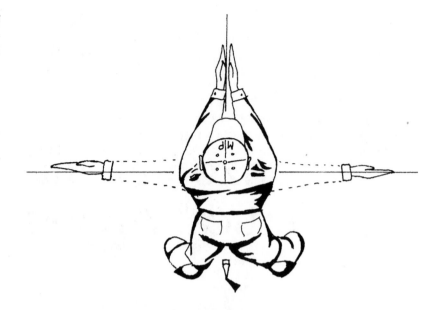

heels together, as if at attention, and close your eyes. Next swing both arms together so that your palms meet. Now open your eyes. Sighting between your thumbs will be approximately 90° to the line you are standing on (see Figure 5-1). Again, practice makes perfect. However, that saying is not entirely true. Only perfect practice makes perfect!

## THE RIGHT ANGLE PRISM

The right angle prism is a more accurate way to lay off a right angle. Again, stand on the point, only this time stand so that the right angle prism is directly over the point. Drop a pebble or hang a bob to be sure you are over the point. Sight through the prism so that the forward and back tangents coincide in the prism. The object seen through the small slot is 90° to the line (see Figure 5-2).

**Figure 5-2** View with right angle prism

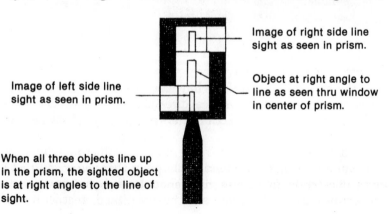

Image of left side line sight as seen in prism.

Image of right side line sight as seen in prism.

Object at right angle to line as seen thru window in center of prism.

When all three objects line up in the prism, the sighted object is at right angles to the line of sight.

You can also use a prism for "bucking in" or "wiggling in" on a line. Use the two prisms to sight along the line and move back and forth until the line is superimposed on both prisms. You are now on a line between the two points.

## THE 3-4-5 TRIANGLE

Probably the most accurate way to lay off a right angle, without fancy equipment, is by the use of a tape and the 3-4-5 triangle. If you have already had trigonometry, this is a simple task. If not, the following shows you how to do it mechanically.

First, measure a multiple of 3 along the line (3, 6, 9, 30, etc.). Next, approximate a right angle with your arms or a prism. Scratch an arc at a multiple of 4 (4, 8, 12, 40, etc.). From the first point, measure a multiple of 5 (5, 10, 15, 50, etc.) and set a point where the 5-distance arc intersects the 4-distance arc. This point will be 90° from the 3 point (see Figure 5-3).

## DIAGONAL CORNERS

When laying out a building, the same principles can be used. Only not all buildings are designed in multiples of 3, 4, and 5. Once the corners have been checked for square and the lengths of the walls are set, the distance between the diagonal corners should be measured. The two diagonals should agree within 0.01 of a foot. If they don't the corners are not square and you have constructed a parallelogram. Use the **Pythagorean Theorem,**

$$a^2 + b^2 = c^2,$$

to calculate the length of the diagonals; or use the length of one wall squared plus the length of the adjoining wall squared equals the length of the diagonal squared. Enter the length of the wall into your calculator and push the $x^2$ key, enter the length of the next wall and push the $x^2$ key. Add the two $x^2$s and push the $\sqrt{x}$ key. This displays the length of the diagonal between the opposite corners.

In Chapter 2, you learned how to set up the transit and the theodolite and turn a single angle. This is really the best way to measure a 90° or any other angle.

For more precise work, the angle may have to be turned several times or repeated ("wound up") to increase the accuracy of the least reading on the vernier. The procedure is the same as you used to double an angle (in Chapter 2), only this time the angle will be doubled three times or "wound up six times." Doubling the angle on a minute transit gives mean of 30 seconds, winding three times gives 15 seconds, four times gives about 8 seconds. Five times is about 4 seconds and the mean of six angles gives nearly a 2-second

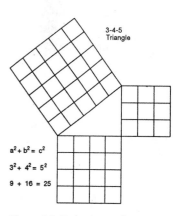

$a^2 + b^2 = c^2$

$3^2 + 4^2 = 5^2$

$9 + 16 = 25$

**Figure 5-3** Pythagorean theorem

accuracy to the measured angle. Some agencies require the angle be wound up twelve times but there is not much use going beyond that. Accuracy begins to drop off because of mechanical and human error.

## TURNING PRECISE ANGLES

When turning precise angles for control surveys or triangulation, greater care must be taken in the setup and handling of the instrument. Make sure that the tripod legs are firmly set in the ground. Sometimes, on shaky ground, it is necessary to drive stakes in the ground and set the legs on them. Also, notice that as you move around the tripod on loose ground, the bubbles of the instrument may move out of adjustment. If this happens, you must find a place for your feet that will allow you to read the instrument and rotate the scope without having to move your feet or shift your weight. Sometimes it requires two persons on opposite sides of the instrument but this is not recommended because everyone reads the vernier a little differently. Check that the plate bubbles are properly adjusted and remain centered when rotated through 360°. For high precision work, it may be necessary to put up an umbrella to shade the instrument from solar heating (see Figure 5-4). Try this experiment when set up in the sun. Hold your hat to shade part of the upper motion and watch the bubble move toward the exposed side as the sun heats the metal and it expands. Think what that does to your accuracy. Also keep your hands off the instrument except when necessary to touch it to make adjustments. Your hands are warm, too.

Do not allow anyone else to move around the setup, especially nonsurveyors. Touch the instrument only on the adjusting knobs and be careful not to try to turn the alidade without loosening the proper motion-locking knob. A pencil can be used to transit the telescope as this will not impart any lateral motion to the scope. For extreme accuracy in angles, the angle should be repeated in each of the four **quadrants** of the circle. The A and B scales can be read and averaged, although these methods are becoming unnecessary because of the precision of modern instruments. It won't hurt to be familiar with them; they might come up on a test.

**Figure 5-4** Solar heating

## REPEATING (WINDING UP) ANGLES

When **repeating (winding up) the angles,** the total of the six angles will sometimes exceed 360°. The number of times the angle has passed 360 can be found by multiplying the first angle by 6 and dividing the result by 360. Multiply the result (rounded off to the nearest whole number) by 360 and add this number to the number six reading before dividing by 6 (see Figure 5-5). The second angle should be read and divided by 2 to check the first reading. If the angles do not check out within 30 sec., start the set over,

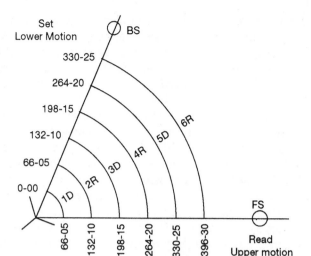

**Figure 5-5** Winding up an angle

as a bad angle has been read on one or two. It is not necessary to record angles 3, 4, and 5. Remember these angles are sets of doubles so 1, 3, and 5 are direct and 2, 4, and 6 are reversed. This will help you keep count by observing the direction of the scope. If the scope is reversed you must be on either angle 2, 4 or 6.

This method of measuring an accurate angle can be used to set an accurate point on the ground. First turn off the desired angle and set a hub and tack on line at the desired distance. Then reset the instrument to zero on the backsight and proceed to wind the angle to the tack 6 times. Find the mean angle and subtract this from the desired angle. The difference is the needed correction. Multiply the tangent of the difference times the distance and move the point the resulting distance in the proper direction (see Figure 5-6).

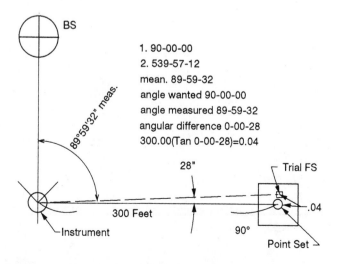

**Figure 5-6** Setting a point by repetitive angles

**VERTICAL ANGLES**

Vertical angles are measured perpendicular to horizontal angles and give us the third dimension necessary to make maps and show differences in elevation. Two methods of measuring vertical angles are used today. One measures the angle in a positive (upward) direction from the horizontal or in a negative (downward) direction from the horizontal (see Figure 5-7).

This method is used on transits that have zero on the vernier when the instrument is level. The other method uses the direction from straight overhead (zenith) to measure angles. Angles under 90° being above the horizon and angles over 90° being below the horizon (see Figure 5-8). On transits, be sure that the level bubble under the scope is leveled and read the vertical circle. If the reading is not zero, check the plate bubble, adjust if necessary, then check the scope bubble again. If there is still a difference, reverse the scope and try again. If the two bubbles still do not agree there is probably an index error in the vertical angles.

Vertical angles should be read direct and reversed to compensate for the index error. Set the center cross hair on the point to

**Figure 5-7** Zero degrees horizontal

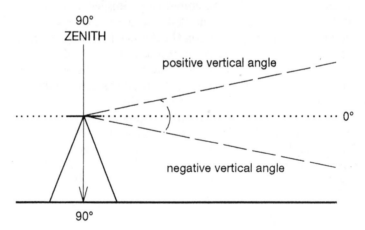

**Figure 5-8** Zero degrees zenith

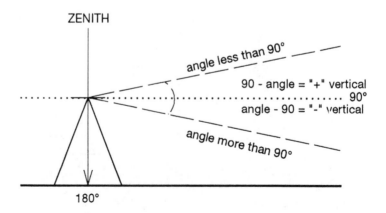

be read and lock the vertical motion-lock screw. Step around to the side of the transit and read the vertical angle. Now release the upper motion lock before transiting the scope. Reverse the scope, set the cross hairs on the same vertical point, and lock the vertical motion. Check the scope to be sure the cross hairs are still on the point. Step around to the side and read the vertical angle again. The mean of the two angles read is the vertical angle and the difference will be equal to the index error. On zenith-reading instruments, the difference between 90° and the reading is the vertical angle. If the angle is less than 90°, subtract it from 90 and the result is the positive vertical angle. If the angle is more than 90°, subtract 90 from the angle and the result is the negative vertical angle.

**DIRECTIONAL INSTRUMENTS**

The instruments discussed so far have been nondirectional, that is, angles can be turned either to the right or left. Angles can be turned in one direction only on directional instruments. There is in effect no lower motion. Angles are read by sighting on the backsight and reading the initial angle (direction). The instrument is then turned to the foresight and the angle (direction) read again. Both angles (directions) are recorded, and the difference between the two readings is the horizontal angle between the foresight and the backsight (see Figure 5-9).

| Position | object observed | DIRECTIONAL THEODOLITE NOTES Observation | Mean D&R | TRIANGULATION Direction | 1 | 2 | 3 | 4 | Mean |
|---|---|---|---|---|---|---|---|---|---|
| Pos. 1 | TS 6 | D 00-00-15 | | | | | | | |
| | | R 180-00-21 | 00-00-18 | 0-00 | 00" | 00" | 00" | 00" | 00" |
| | TS 5 | D 50-04-31 | | | | | | | |
| | | R 230-04-39 | 50-04-35 | 50-04 | 17" | 15" | 20" | 18" | 18" |
| | TS 7 | D 89-09-40 | | | | | | | |
| | | R 269-09-44 | 89-09-42 | 89-09 | 24"07" | 05" | 10" | 09" | 08" |
| | TP 9 | D 134-07-16 | | | | | | | |
| | | R 214-07-21 | 134-07-18 | 134-07 | 00" | 56" | 02" | 02" | 00" |
| | TP 10 | D 181-17-21 | | | | | | | |
| | | R 01-17-23 | 181-17-22 | 181-17 | 04" | 00"-58 | 06" | 09" | 05" 04" |
| Pos. 2 | TP 6 | D 45-02-30 | | | | | | | |
| | | R 220-02-37 | 45-02-37 | 0-00-00 | | | | | |
| | TP 5 | D 95-06-47 | | | | | | | |
| | | R 275-06-51 | 95-06-49 | 50-04-15 | | | | | |
| | TP 7 | D 134-11-38 | | | | | | | |
| | | R 314-11-40 | 134-11-39 | 89-09-05 | | | | | |
| | TP 9 | D 179-09-30 | | | | | | | |
| | | R 359-09-30 | 179-09-30 | 134-06-56 | | | | | |
| | TP 10 | D 226-19-34 | | | | | | | |
| | | R 46-19-35 | 226-19-34 | 181-17-00 | | | | | |
| Pos.1 Reobservation | | | | | | | | | |
| | TP 6 | D 00-00-05 | | | | | | | |
| | | R 180-00-08 | 00-00-07 | | | | | | |
| | TP 7 | D 89-09-13 | | | | | | | |
| | | R 269-09-15 | 89-09-14 | 89-09-07 | | | | | |

**Figure 5-9** Directional theodolite notes

**Figure 5-10** Zeiss total stations. *Courtesy of Carl Zeiss, Oberkochen, Germany*

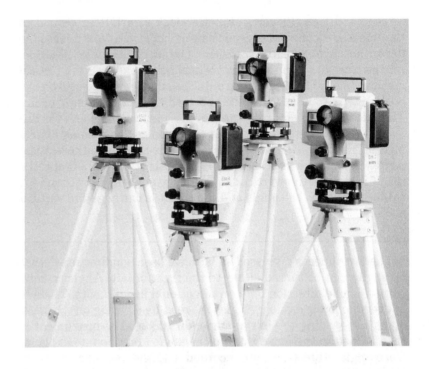

## DIGITAL TOTAL STATIONS

Now, the good news. There are instruments on the market today that correct for all these things and they do it automatically for us.

Electronic Theodolites that have digital readouts of angle and distance are now available through all the major instrument manufacturers. Instruments such as the Wild T1600, the Zeiss ETH3, and the Topcon ETL-1 have automatic corrections for circle eccentricity (see Figure 5-10). They also have two axis compensators for determination of the vertical axis tilt. The horizontal and vertical angles are read directly off a display screen to the nearest 1 to 3 seconds. Zero can be set by merely touching a button after the cross hairs are set on the target. Automatic averaging of multiple sightings can be done with a display of the accuracy.

## WIGGLING IN ON LINE

On occasion you will not be able to see from point-to-point with the instrument. If an intermediate point can be found from which it is possible to see both points, a procedure known as wiggling in on line can be used. If you have a right angle prism, use it to get an approximate line point. Set up the instrument on the point found and sight the backsight. Now transit the telescope, **throw over** or **plunge the scope** to see if the foresight is in the field of view. If the foresight can be seen in the scope estimate how far it is off line. Now estimate how far the setup point is along the line

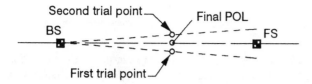

Second trial point — Final POL

BS

First trial point —

FS

**Figure 5-11** "Wiggle in"

between the backsight and the foresight (see Figure 5-11). This does not have to be accurate, just "guesstimate" the distance. Now, move the instrument over a proportional distance in the direction of the foresight. If you missed by 1 ft. and you are setup about one third of the distance to the foresight, move the instrument over about three tenths of a foot. Now repeat the whole procedure. With practice, you should be able to **buck in** or **wiggle in** in about three or four tries. Also, do not forget to double the angle once you think you are close. Just reverse the scope, using the lower motion, and repeat the throw over to check for vertical tilt in the setup.

Sometimes a "double wiggle in" is necessary. This should be done using two instruments, but the procedure is the same. If you have a place to practice these, you should give it a try. It can be fun once you have overcome the aggravation of going the wrong way or too far just when you think you have got it.

## Review Questions

1. A building is laid out and stakes are set at the four corners. The length of the building is 70 feet 8 inches and the width is 38 feet 6 inches. What is the diagonal?

2. An angle to the left was measured with a 1-min. transit. The following angles were recorded:
   1. 67°27′
   2. 134°54′
   3. 44°43′
What is the mean angle?

3. If the angle in question 2 was being used to set a point at 300.00 ft. from the instrument and the required angle was 67°27′ to the right, how much, and in which direction, would you move the point being set?

4. If the initial reading on a directional instrument was 00°14′26.3″ and an angle of 174°18′15.2″ was read to the next point, what is the horizontal angle?

5. Set up a transit over a point and pick at least five points around the horizon. Turn the angle to each point from the preceding point. The total of the angles turned should equal 360°. How much error did you have?

6. Now wind up each angle in question 5 six times and calculate the means. Add the means and see if the difference from 360° was greater or lower. By how much?

7. Repeat questions 5 and 6 with a theodolite and see if the angles were more precise. If they were, what was the reason?

8. Set up a short closed traverse of at least seven points. Set up on each point and turn the angle between the foresight and the backsight. Add up the angles and calculate the closure. For interior angles, the number of points minus 2 times 180 should equal the number of degrees in the traverse: $n - 2 = (180)$ interior degrees.

9. Set two hubs about 300 ft. apart. Using an instrument (transit or theodolite), set a **point on line** (POL) between the two hubs by the "wiggle in" method. Check the POL by setting on one of the original hubs and sighting the other. How far were you off line?

10. Using a transit and a tape, set a point exactly 90° from one of the hubs in question 9.

# Azimuths and Bearings—The Direction of a Line

**6**

## Objectives

After completing this chapter, the student should be able to:

1. Distinguish between astronomic north and magnetic north.
2. Compare the advantages and disadvantages between using azimuth or bearings to define the directions of traverse lines.
3. Calculate the bearings/azimuths of a 5-sided traverse given the interior angles and a basis of bearing.
4. Relate the method used to obtain an astronomical bearing using:
   A. the sun;
   B. Polaris.
5. Convert from magnetic bearings or azimuths to astronomic bearings or azimuths and vice versa.
6. Determine the change in magnetic declination for a gain in place and time.
7. Convert from bearings to azimuths and vice versa.

The earliest people needed a directional reference to find their way back to the cave. People needed a more universal method than using landmarks as they traveled farther from home. The sun and stars were soon discovered to be constant beacons to the more knowledgeable traveler.

**Figure 6-1** Old compass

## THE MAGNETIC NEEDLE

About the tenth century, the Chinese began to use the magnetic needle to tell direction. By 1269, Peregrinus described a pivoting compass with four quadrants of 90° and the cardinal points of north, south, east, and west. The earliest surveys in America were made using a compass (see Figure 6-1). Directions were stated as being so many degrees east or west of the cardinal north or south points that the needle seemed to point to so accurately.

## DECLINATION

It was known to a learned few that the needle did not always point to **true north**, the direction of the pole star, or **Polaris**. This discrepancy could be caused by local magnetic attraction or by variations in the Earth's magnetic field called **declination** (see Figure 6-2). It caused the needle to point in directions other than

**Figure 6-2** True north and magnetic north

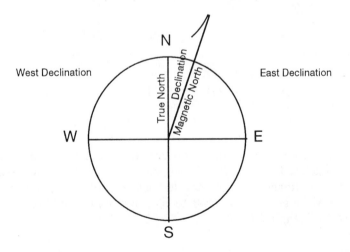

## Magnetic Declination in the United States — 1985

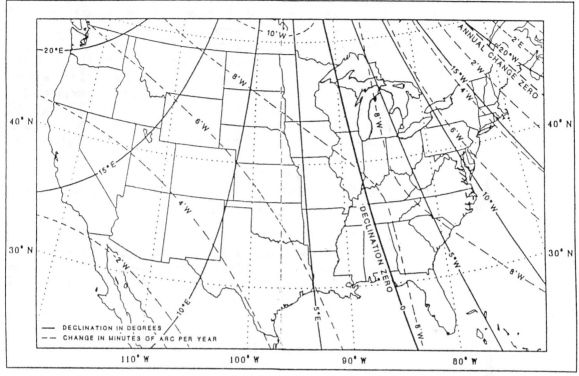

**Figure 6-3** Isogonic chart. *Courtesy of the U.S. Corps of Engineers*

**magnetic north.** It was known that the north magnetic pole was not in the same position as the true pole. It even moves about so that the lines of declination change from year to year. The annual change can be found on an **isogonic chart** (see Figure 6-3).

## IN THE FOOTSTEPS OF THE ORIGINAL SURVEYOR

If it becomes necessary for you to retrace an old survey that was originally run "by the needle," you will need to find a method to correct your **bearings** to the original lines as run. If you can find two of the original points for which you have the old bearing, a new bearing can be taken and the difference found by subtracting one from the other. This difference can then be applied to the other line, for which you have no original points, to reconstruct the direction and set new points. Remember when retracing a survey: ALWAYS WALK IN THE FOOTSTEPS OF THE ORIGINAL SURVEYOR! This means that you must correct your chain and compass to his or her chain and compass, not theirs to yours. If the original surveyor did not record the declination that was used, it is possible to recreate it by using the annual change from the isogonic chart. Try to locate the chart that was in use at the time of the original survey and use the declination given. If a copy of the old isogonic chart cannot be found, use a current chart. Multiply the annual change by the number of years since the original survey was made and the chart date. The use of recent charts to go back too far can be misleading. For instance, the isogonic chart for 1885 gives the declination at Roswell, New Mexico, as 12°E. The 1985 chart gives the declination as 10°E and the annual change as 4'W. 1985–1885 = 100 years times 4 minutes equals 400 minutes or 6°40' change: 10° + 6°40' = 16°40' as the calculated declination for 1885. However, the 1885 chart shows 12°, which amounts to a 4° 40' discrepancy. Which do you use? In this case, you should try to find a part of the original survey to help you decide. If no trace of the original survey remains, use your best judgment to decide, but be sure to record in the notes which you used and your reason for doing so.

## AZIMUTHS

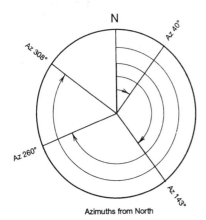

Figure 6-4 Azimuths

It is easier to use azimuths than to use bearings to convert between magnetic and true direction. If the declination is east, then magnetic azimuths will be less than true. If the declination is west, then magnetic is greater than true.

> East is less: ADD the declination to magnetic to get true.
> West is more: SUBTRACT the declination from magnetic azimuth to get the true azimuth.

An easy way to remember this: EAST IS LEAST—WEST IS BEST! Just reverse the procedure to go from true to magnetic.

Azimuths are lines of direction measured in a clockwise direction from north, for plane surveying, and from south, for geodetic surveys (see Figure 6-4). The directions are measured from a meridian, which is a line of longitude that passes through both the north and south poles. Meridians converge at the poles and are farthest apart at the equator. The azimuth of a line will be between 0° and 360°. Using north azimuth is common practice for local surveyors. When north azimuth is used, the numerical values of azimuths and bearings in the first quadrant are the same (see Figure 6-5). In the second quadrant, subtract the azimuth from 180° to get the bearing. In the third quadrant, subtract 180° from the azimuth to obtain the bearing. In the fourth quadrant, subtract the azimuth from 360° to get the bearing.

The back azimuth of a line is the reverse direction of the line. If the azimuth of a line is under 180°, adding 180° to the forward azimuth gives the back azimuth. If the azimuth is over 180°, subtracting 180° from the forward azimuth will give you the back azimuth. Back azimuths are used in many computer programs as a direction away from the starting point in traverse calculations.

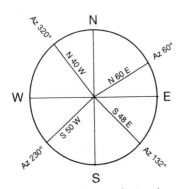

Figure 6-5 Bearings and azimuths

The azimuths between control monuments, the azimuth marks set by government agencies, may be south azimuth. So, if you wish to use them to find a bearing, be sure to use the proper direction of the line. Read the reference notes to be sure what you are given.

## THE BEARING OF A LINE

The bearing of a line is the angle measured from north or south in either an east or a west direction (see Figure 6-6). The first letter gives the origin (north or south), the numbers give the degrees, minutes, and seconds of the arc the line has rotated through, and the final letter gives the direction of **bearing rotation** (east or west). A bearing of N 36-24-15 E (surveyor shorthand for 36°24′15″) begins at north, turns through an angle of 36 degrees 24 min. 15 sec. in an easterly direction (see Figure 6-7).

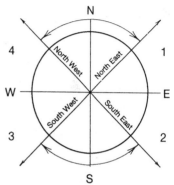

**Figure 6-6** Bearings

As bearings always begin at north or south and rotate either east or west it is obvious that they cannot exceed an angle of 90°. The back bearing of a line is the same angle as the forward bearing but with the opposite quadrant's direction. The back bearing of N 45-00-00 E is S 45-00-00 W and vice versa. To change from bearings to azimuth in the north-east quadrant, just drop the letters. The angles are the same. In the south-east quadrant, subtract the bearing from 180° to get the azimuth. In the south-west quadrant, add the bearing to 180° to get the azimuth. In the north-west quadrant, subtract the bearing from 360° to get the azimuth.

When calculating the angle between two bearings, they may be converted into azimuths and the difference found by subtraction (see Figure 6-8). Another method is to place both lines in the same **hemisphere** by using back bearings, if necessary (Figure 6-9), and, if both bearings are in the same quadrant, subtract them from each other. If the bearings are in adjacent quadrants, add them together. The results can be added to or subtracted from 180° to get the needed angle.

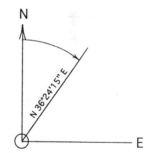

**Figure 6-7** Bearing rotation

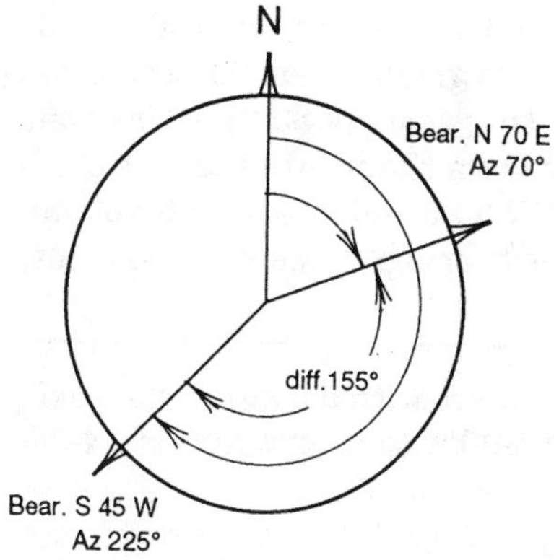

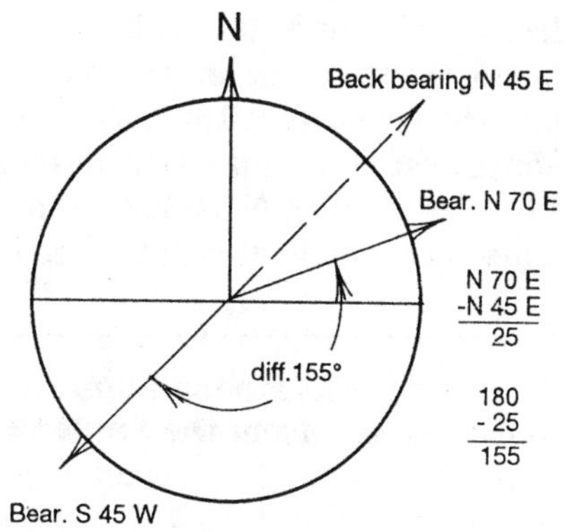

**Figure 6-8 (Left)** Difference in azimuths

**Figure 6-9 (Right)** Difference in bearings

**Figure 6-10** Bearing traverse

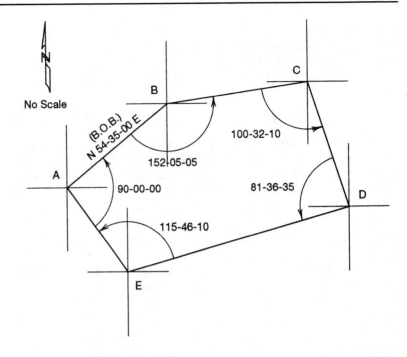

## A BEARING TRAVERSE

The bearings around Figure 6-10 start with the BOB (basis of bearing) of N 54-35 E between points A and B and continue clockwise around the traverse. The bearing of B to C can be calculated by subtracting 152-05-05 from 180° and adding the result (27-54-55 to 54-35-00) to get 82-29-55. Because this is not over 90°, the resulting bearing is N 82-29-55 E. Subtracting 82-29-55 from the angle at point C of 100-32-10 gives an angle of 18-02-15 or S 18-02-15 E. The bearing of N 82-29-55 E is changed to S 82-29-55 W or the reciprocal bearing. Course C to D is in the southeast quadrant. Therefore, if the angle in the southwest is subtracted from the interior angle of 100-32-10, the remainder of 18-02-15 must be in the southeast quadrant (or S 18-02-15 E). If the back bearing from D to C of N 18-02-15 W is added to the interior angle of 81-36-35, the resulting angle of 99-38-50 is greater than 90° and must be subtracted from 180°. The resulting angle of 80-21-10 is in the southwest quadrant so the bearing from the point D to E is S 80-21-10 W. Adding N 54-35-00 E and N 35-25-00 W equals 90-00-00. This is the angle at point A, so our bearings close with no error.

## AN AZIMUTH TRAVERSE

The azimuths around Figure 6-11 start with changing the basis of bearing from a bearing of N 54-35-00 E to an azimuth of 54-35-

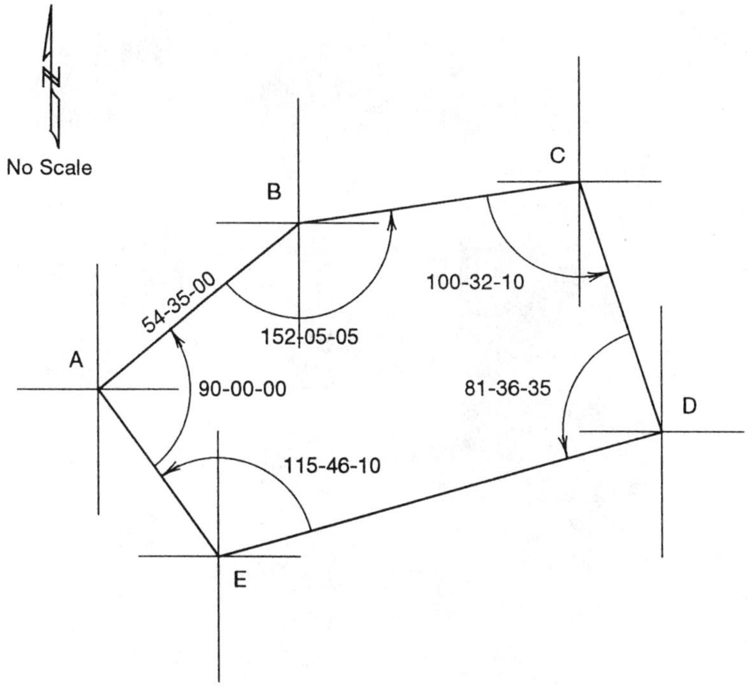

**Figure 6-11** Azimuth traverse

00. Adding 180° to the forward azimuth of 54-35-00 gives a back azimuth of 234-35-00. Subtracting the interior angle of 152-05-05 gives a forward azimuth from B to C of 82-29-55. Adding 180° to the forward azimuth gives a back azimuth of 262-29-55. Subtracting the angle of 100-32-10 from the back azimuth gives an azimuth of 161-57-45 from C to D. Adding 180° to 161-57-45 gives the back bearing D to C. Subtracting 81-36-35 from 341-57-45 gives 260-21-10 D to E and 260-21-10 minus 115-46-10 equals 144-35-00 or the back azimuth A to E. Adding 180° to the back azimuth gives a forward azimuth from E to A of 324-35-00. Now subtract 90-00-00, the interior angle at A, from the back azimuth of 144-35-00. The answer of 54-35-00 is the check to the forward azimuth of the starting line of A to B. If you calculate the azimuths in a counter-clockwise direction, the mathematics are much simplified.

**TAKING ASTRONOMICAL AZIMUTHS**

With the coming of the pocket programmable calculator (see Figure 6-12) and the laptop computer, the ease of taking astronomical azimuths is causing an increase in the use of astronomical north as a basis of bearing. Accuracies of 10 arc sec. are attainable using the hour angle method. This will give a positional accuracy of 0.25 ft. in a mile.

**Figure 6-12** Hewlett-Packard 48SX expandable calculator. *Courtesy of Hewlett-Packard Company*

## THE HOUR ANGLE METHOD

The additional equipment needed for a celestial observation using the hour angle method is simple:

1. An accurate timekeeping device, such as a digital stopwatch or a time module, on your calculator. An ephemeris is not necessary, if you use one of the programs for the calculator that calculates the ephemeris data.
2. A sun filter is necessary if you are using a total station and is needed for taking direct observations of the sun. Remember, NEVER LOOK DIRECTLY AT THE SUN THROUGH AN INSTRUMENT WITHOUT AN APPROVED SOLAR FILTER ATTACHED TO THE SCOPE! Severe eye damage will result if you do so without proper protection.
3. A time receiving portable radio, such as a Radio Shack TimeKube, is inexpensive and dependable. The National Bureau of Standards Radio Stations broadcast on frequencies of 2.5, 5, 10, 15, and 20 MHz. The U.S. Naval Observatory's "Master Clock" can be reached by dialing 900-410-TIME. WWV can be dialed at 303-499-7111.

Time is announced in Universal and Eastern Standard (Daylight) time. The minute will be announced by a voice just before the 59th second is skipped. An 800-millisecond tone will announce the minute. The tone will be followed by clicks each second with the 29th and 59th skipped. The time announced is in "Coordinated Universal Time" (UTC). UTC is the time in Greenwich, England, at the Royal Observatory, which is at 0° longitude. The time will be 5 hours greater than Eastern time, 6 hours greater than Central time, 7 hours greater than Mountain time, and 8 hours greater than Pacific time. If you hear double clicks for the 1st, 2nd, 3rd, or 4th seconds of minute, the DUT1 (Universal Time) correction for your time zone is plus 0.1 seconds times the number of ticks. If no double clicks are heard through the first 8 seconds, but are heard on the 9th, 10th, 11th, or 12th seconds, the DUT1 correction is minus 0.1 seconds times the number of ticks. The DUT1 correction is added or subtracted from the UTC time to get UT1.

**COORDINATED UNIVERSAL TIME**

An error of 1 sec. in time can cause an error in azimuth of 15 sec. so be as careful as possible in checking the time. Start the stopwatch as close as possible to 0 sec. on the radio and stop it when the vertical cross hair is tangent to the leading or trailing edge of the sun. The leading and trailing limbs should be shot alternately to allow for correction to the center of the sun for calculation purposes. The elapsed time in seconds is then added to the UT1 time to get the UT1 time of observation.

The azimuth is turned from some permanent backsight to the limb of the sun being shot and is read after the time has been recorded. The latitude and longitude of the station, where the instrument is set up, can be scaled from a 7.5-min. USGS quad sheet. The distances can be scaled accurately by using proportional parts, as in topo contours, to find the number of seconds to be added to the map tics for both latitude and longitude.

Geodetic and astronomic positions are not exactly the same due to deflections from the vertical caused by variations in the pull of gravity. These variations are caused by the proximity of large masses, such as mountains, and are called the **LaPlace corrections**. The corrections can be found in tables published by the NGS (National Geodetic Survey).

The astronomic azimuth should finally be corrected to grid azimuth to make it conform to state plane coordinates. The correction can easily be made using a computer with the proper software based on NAD 83 datum.

One method of taking a solar observation is: level the instrument carefully. *Be sure that a solar filter is installed.* Take an initial

**THE LEADING OR TRAILING EDGE OF THE SUN**

sighting on the backsight with the instrument direct. Take three sightings on the sun with the instrument in the direct position. Reverse the scope and take three sightings on the sun with the instrument reversed. Take a sight on the backsight target with the instrument reversed. The recorded data may be reduced in the field or the office and the true bearing from the station to the reference mark can be calculated and recorded.

## THE NORTH STAR

Polaris is similar to shooting the sun, however it is easier to shoot Polaris at night. Polaris, or the North Star, can be seen during daylight hours but the procedure is more difficult. Polaris is easier to find after dark when the Big Dipper is visible. The two stars on the outside of the cup of the Big Dipper point to Polaris. Its distance is five times greater than the distance between the two stars (see Figure 6-13).

Polaris is not exactly on the north celestial pole, but it makes a small circle around it every 24 hours (see Figure 6-14). When Polaris is at the top of the circle, it is said to be at upper culmination. When at the bottom of the circle, it is at lower culmination. When Polaris is at the observers right side of the circle, it is at easterly elongation, and on the left side, at westerly elongation. The radius of the circle is only about 54 min. of angle. The amount of arc between the north celestial pole and Polaris is known as the polar distance and can be obtained from the ephemeris tables by subtracting the declination of Polaris from 90°.

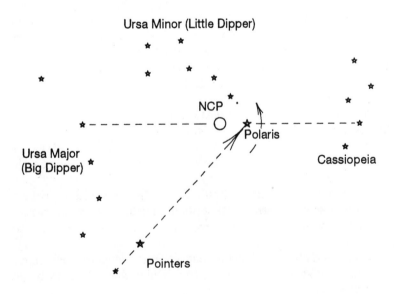

**Figure 6-13** Locating the North Star (Polaris)

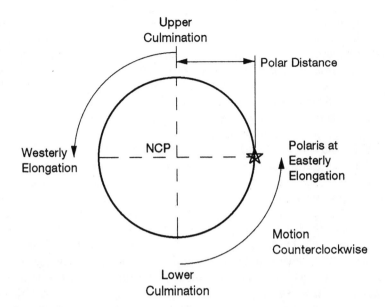

**Figure 6-14** Upper and lower culmination and east and west elongation

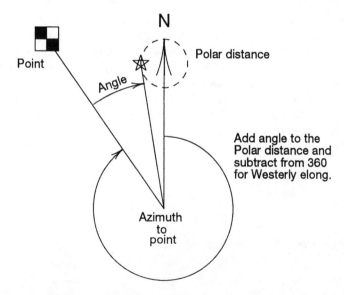

**Figure 6-15** Finding azimuth of point from Polaris at elongation

The easiest time to observe Polaris for azimuth or astronomical north is at easterly or westerly elongation. At elongation, the star appears to be traveling straight up at easterly and straight down at westerly. This allows time for repeated sightings without the necessity of keeping good time for the sightings. The polar distance can then be set off on the mark to give true north (see Figure 6-15).

**NIGHT OBSERVATIONS**

Night observations require the use of illumination for the cross hairs and circles on the instrument and a flashlight to illuminate the reference mark. For observations other than at elongation, accurate time will need to be kept of the pointings. Because of the slow apparent movement of Polaris, time is not as critical as with a solar observation. At culmination Polaris only moves 18 sec. of arc horizontally for every minute of time and only 12 sec. of arc vertically per minute of elongation.

The instrument must be set up over a point and carefully leveled before beginning observations of Polaris. A reference mark (hub) should be set as far away from the instrument point as practical for night sighting. This point should be as close to north as you can get so that the correction offset will not be more than a few tenths. Sight the instrument on the reference mark with 0° on the horizontal circle. Next, sight the instrument on Polaris and the plate bubble checked for level and adjusted, if necessary. Now center the vertical cross hair on the star and mark the time of sighting. Reverse the scope and check the bubble again. Sight on the reference mark again, just the same as winding up an angle, using the lower motion. Sight on the star again and check the bubble. Set the vertical cross hair on the star and mark the time. The difference between the two angles should be only a few seconds. If it is too large, rerun the set of angles. Repeat the procedure for three more sets. If the angles check, the field notes can be used to calculate the amount of arc between the reference mark and true north. Use the sine function of the mean angle, times the distance to the reference point from the instrument point, to calculate the offset to true north. The offset can be set off in the proper direction using a tape. If the offset falls off the hub, but close to it, be careful not to disturb the reference hub when driving a new hub.

## Review Questions

1. The interior angles in a counter clockwise traverse are:

   P1. 83-15-22 angle right

   P2. 192-45-58 angle right

   P3. 96-18-42 angle right

   P4. 266-02-25 angle right

   P5. 84-50-14 angle right

   P6. 92-55-49 angle right

   P7. 153-48-33 angle right

P8. 110-02-57 angle right

The azimuth of line 6 to 7 is 89-38-19. What are the azimuths of the other lines?

2. What are the bearings of the lines in question 1?

3. What are the interior angles in the following traverse?

    1 – 2 N 47-28-00 E

    2 – 3 S 8-27-30 W

    3 – 4 S 56-27-00 W

    4 – 5 N 26-26-30 E

    5 – 6 N 39-18-00 W

    6 – 1 S 80-20-30 E

4. How many hours must be added to UTC for your time zone?

5. If you hear double clicks in the 1st and 2nd seconds of the minute, how much time must be added to the UTC time?

6. Why do you need to shoot the leading and trailing limbs of the sun on the hour angle method?

7. How do you determine latitude and longitude of the instrument station?

8. What is the LaPlace correction?

9. Astronomic azimuths should be recorded to what azimuth for use in state plane coordinates?

10. Why should you use an approved solar filter on the instrument?

11. How many sets of direct and reverse sightings should be taken for an astronomical shot for azimuth?

12. How do you locate Polaris?

13. Polaris travels in a small circle around the celestial pole. As it progresses, as viewed through the scope, the points farthest from the pole are called what?

14. At the top and bottom of the circle, Polaris is at what position(s)?

15. What is the distance from the north celestial pole to Polaris called?

# Traverse Methods and Calculations

## Objectives

After completing this chapter and the field practice, the student should be able to:

1. Distinguish between an open and a closed traverse.
2. Using a transit/theodolite and tape or EDM, turn all angles and measure all distances on a traverse of at least five sides.
3. Use the proper note form to record the field data for a traverse of at least five sides.
4. Calculate the angular closure of a closed traverse of at least five sides.
5. Adjust the angular closure of a closed traverse of at least five sides.
6. Calculate the latitudes and departures of a closed traverse of at least five sides.
7. Using the compass rule, adjust the latitudes and departures of a closed traverse.
8. Calculate the local coordinates for a closed and adjusted traverse.
9. Inverse between given coordinates of two points and find the bearing and distance between the coordinated points.
10. Calculate the area of a closed traverse using the coordinates of the corners.

In the previous chapters, you have learned to use survey equipment and methods to measure distances and angles. Now you can put all that knowledge to use in the basic field survey: the **traverse**. A traverse is a series of lines, of measured length, connected by measured angles.

**THE TRAVERSE**

**OPEN AND CLOSED TRAVERSES**

There are two kinds of traverses: open and closed. An **open traverse** begins at a known point and goes to another point whose location is uncertain. This point cannot be checked for error in measurement or calculation, except by another traverse. An open traverse can be tightened up a little by doubling all angles and double taping all distances (see Figure 7-1). However, you must never use it in a control survey.

A **closed traverse** is exactly that. It begins on a point of known position and closes to another point of known position (see Figure 7-2). A closed traverse can be checked for accuracy mathematically and any minor error can be adjusted out. Closed traverses can close back to the starting point or close to another point of known position, as in a **connecting traverse**.

We use a traverse to find the relative position of existing points, to set points for a boundary, to set control for a project, or to connect two or more previous traverses. A connecting traverse is the most reliable for distance and direction because it cannot have a rotation. A traverse that closes on itself can be rotated and still close mathematically.

**Figure 7-1** Open traverse

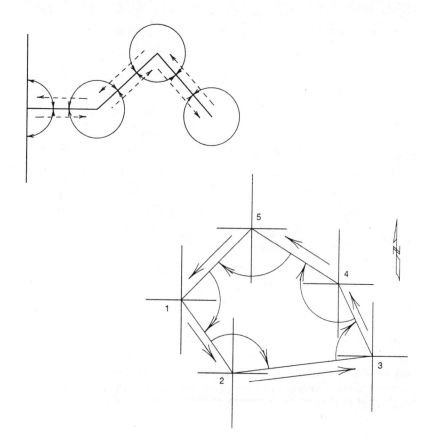

**Figure 7-2** Closed traverse

The field angles used in a traverse are of three types; angles right, angles left, and deflection angles. **Deflection angles** may be either right or left. Be sure to show clearly in the notes the type of angle used (see Figure 7-3).

Traverse angles are of three types: interior angles, exterior angles and, deflection angles (see Figure 7-4). Distances are measured with tapes or EDM equipment. All distances should always be measured twice and averaged. Reciprocal measurements, forward and back, are also used. When measuring slope distances with the tape, be sure to get two separate measurements (high and low) with the scope direct and reversed. Get direct and reversed vertical angles with the EDM and read the distance in both feet and meters for a check. All distances should be reduced to horizontal and compared before moving the instrument ahead.

Angles should be wound up six times. Angle number 1 should be compared to the mean of the six angles before the instrument is moved ahead. If the comparisons show too much spread for the order of the traverse, the angles and distances should be measured again.

**FIELD ANGLES**

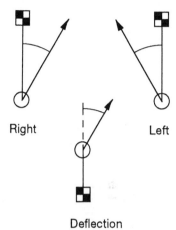

Right          Left

Deflection

**Figure 7-3** Types of horizontal angles

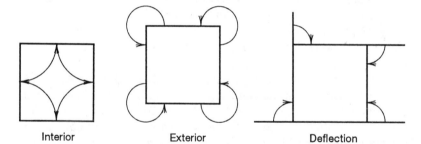

Interior          Exterior          Deflection

**Figure 7-4** Types of traverse angles

**A FIVE-SIDED TRAVERSE**

Figure 7-5 shows a five-sided traverse. The points are shown consecutively, A through E, in a clockwise direction. To measure the interior angles of this traverse, the angles should be turned to the left, starting at point A and backsight point E. The angles can be turned to the right if point B is used for the backsight and the angle turned to point E. If the instrument is set up on point A, and point E is sighted with the scope inverted and then thrown over so that it is pointing away from E along the extension of line E-A, a deflection angle to the right can be turned to point B.

The choice of which method to use depends on the type of instrument in use and the conditions found on the job site. With a directional instrument, it would be necessary to do an exterior angle traverse to turn all angles to the right or to do the traverse in a counterclockwise direction. Because of this, it is common to set up a traverse for a directional instrument to be run counterclockwise.

**Figure 7-5** Five-sided traverse

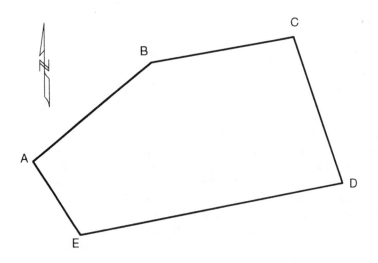

**AN INTERIOR ANGLE TRAVERSE**

Go through the steps to do an interior angle traverse of Figure 7-5 using a nondirectional instrument and taping the distances. Before starting the survey operations, the boundary of the job should be walked and all points of record found and marked. In your survey you will be lucky and find all the corners just where the owner said they would be. This does not happen very often, so be sure to check.

A sight should be set on point E, facing toward point A. Check to be sure that it can be seen from point A. Set up the instrument on point A, as the tapeperson goes ahead and sets a sight on point B. Check through the scope to be sure both sights are visible. Sight the instrument on point E with 0-00-00 locked in the upper motion. Lock the lower motion and fine adjust, using the lower motion-tangent screw. Release the upper motion lock and rotate the scope to point B. Sight on point B, using the upper motion tangent lock and screw. Now read the angle. Be sure to read the correct horizontal angle. Remember, the angle was turned to the left. Record the angle as number 1 (see Figure 7-6) and release the lower motion lock. Reverse the telescope and sight on point E again, using the lower motion.

One of the most common errors in turning angles is moving the wrong tangent screw. Notice that the upper and lower tangent screws have different shapes to help you avoid this. Learn their shapes by feel. Release the upper motion lock and turn the scope to sight on point B, lock the upper motion and fine adjust with the upper motion-tangent screw. Now read and record the doubled angle as number 2. Before going any farther with turning angles, check the second angle by dividing it by two and comparing the answer with number 1. If the difference is minimal, go ahead and

| 5 Sided Traverse 7-6 | pg. 1 of 2 | | | | | pg. 2 of 2 |
|---|---|---|---|---|---|---|
| | | | Dist. C - D | | | |
| ⊼ @ "A" B.S "E" F.S. "B" ∡ Lt. | | | | | | |
| | | ⊼ @ "D" B.S "C" F.S. "E" ∡ Lt. | | | |
| 1 - 90-00-00 | | | | | | |
| 2 - 180-00-00 | | 1 - 81-36-35 | | | | |
| 6 - 180-00-00 | | 2 - 163-13-10 | | | | |
| M - 90-00-00 | | 6 - 129-39-30 | | | | |
| | | M - 81-36-35 | | | | |
| Dist. A - B | | | | | | |
| | | Dist. D - E | | | | |
| ⊼ @ "B" B.S "A" F.S. "C" ∡ Lt. | | | | | | |
| | | ⊼ @ "E" B.S "D" F.S. "A" ∡ Lt. | | | |
| 1 - 152-05-05 | | | | | | |
| 2 - 304-10-10 | | 1 - 115-46-10 | | | | |
| 6 - 192-30-30 | | 2 - 231-32-20 | | | | |
| M - 152-05-05 | | 6 - 334-37-00 | | | | |
| | | M - 115-46-10 | | | | |
| Dist. B - C | | | | | | |
| | | Dist. E - A | | | | |
| ⊼ @ "C" B.S "B" F.S. "D" ∡ Lt. | | | | | | |
| 1 - 100-32-10 | | 90-00-00 | | | | |
| 2 - 201-04-20 | | 152-05-05 | | | | |
| 6 - 243-13-00 | | 100-32-10 | | | | |
| M - 100-32-00 | | 81-36-35 | | | | |
| | | 115-46-10 | | | | |
| | | 540-00-00 | | | | |

**Figure 7-6** Field notes for a five-sided traverse

wind up the rest of the angles. Only angle number 6 needs to be recorded. If the difference between the first and second angles is greater than the minimum reading of the vernier, you would be better off to start all over again.

The whole process of setting up the instrument and reading six angles should not take more than 10 or 12 minutes. After the sixth angle has been read and recorded, divide it by 6 to get the mean. As the angle measured was 90° and 360° was passed on the fourth turn, the sixth angle would be recorded as 180°. Add the 360° to the 180° to get 540° and then divide that by 6 to get 90°. The mean should be within a couple of seconds of the expected angle. If everything checks out all right, the instrument can be moved to point B while the tapeperson sets a sight on point A. After the instrument is set up on point B, the distance from point A to point B can be taped and the distance recorded in the field book.

The distance can be measured when the instrument is set up on A, if you chose to do so. If the distance is to be double taped, you might want to measure A to B with the instrument set at B. The tapepeople go ahead to point C and set the sight and then measure back to B.

## TYPES OF SIGHTS

There are several types of sights that can be set on a tape traverse (see Figures 7-7, A-E). Make sure that any type of sight that you choose is directly over the point. If the sight is plumbed over

**Figure 7-7** Types of sights: A) nail in target; B) nail in target in stake set behind point; C) crows foot on lath set behind point; D) target in cut stake set behind point; E) error in alignment due to offset sight. *Courtesy of Michael Perniciaro*

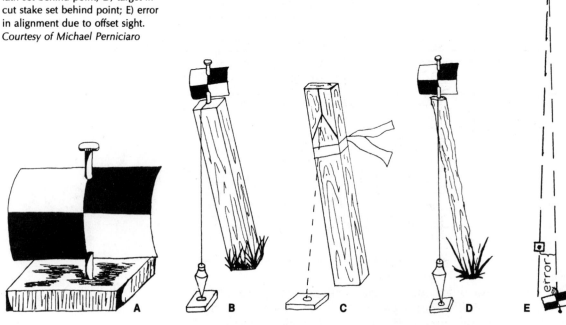

the cup-tack, the alignment of the sight is not critical. If the sight is set behind the point, it must be lined up with the instrument; if not, an alignment offset will occur if the target is off the prolongation of the line through the point to the instrument.

After measuring the distance from C to B and turning and recording the angle, move the instrument ahead to C and set up. A sight is set at B and the line measured ahead from B to C for the second measurement of B to C. The tapepeople then go ahead to D and repeat the process.

On long lines, through hilly country, it may be necessary to set intermediate points on line. If a rise in the ground between C and D prevents seeing the sight on D, it may be possible to set a "wiggle-in" point on the high ground between the two points and, from that point, give line for taping.

A sight can be left on the intermediate point to use for a backsight. If the vegetation or terrain prevents seeing the foresight, it may be necessary to run a traverse between the points. Remember that each of the traverse points becomes another point in the main boundary traverse and must be included in the adjustment of the traverse.

Try to balance the length of the legs of the subtraverse as much as possible. Many out-of-balance short sights will ruin the accuracy of the whole traverse. This leg might receive more adjustment on the closure adjustment due to increased chance of error in multiple setup lines over single setup lines. The errors increase as the number of opportunities for error increase.

If a sight is left on point A facing point E, it will save a trip ahead to set a sight when point E is occupied. Sometimes a little planning can save a lot of walking. Your ability to think and plan is your greatest work- and time-saving aid in the field.

## AN ANGULAR CLOSURE

Check the angular closure of the traverse before leaving the job. On an interior angle traverse, such as this one, the number of degrees in the closed figure should add up to $n - 2$ (180); in other words, the number of points in the traverse (n) minus 2, times 180°. Three times 180° equals 540° in the inside angles of our traverse. If your answer is off more than a minute or a few seconds, and if control must be tight, you may be able to re-turn 1 or more of the angles that you feel might be in error. It could save another trip to the job site.

When you arrive back at the office, make copies of the notes and give the originals to the person that will do the closure calculations and adjustment. If there are any questions about the job, they should be settled before you leave. That way computations can be done without having to talk to the party chief the next day when the party is in the field. Closing and adjusting this traverse is discussed later in the chapter (under Traverse Calculations); but first an EDM traverse using deflection angles will be studied.

## THE CONNECTING TRAVERSE

The connecting traverse shown in Figure 7-8 is typical of a preliminary survey used for road design. The job is to run a line from a point on the center line *(CL)* of Oahu Road 550 feet north of the monument at the *CL* intersection of Oahu and Hilo Road, to a point on the *CL* of Maui Road 500 feet north of the monument at Maui and Hilo Road. The basis of bearing will be the *CL* of Oahu Road, which is N 9-15-20 E. The recorded *CL* bearing of Maui Road is N 4-36-18 W. Set a spike and tin in the pavement on the *CL* of Oahu Road as point of beginning (POB) and give it station 0+00. Stations are even hundreds of feet measured in the direction of travel. A 1+00 would be the same as 100 feet.

To begin the survey, recover the monuments at the intersections of Hilo and Oahu and of Maui and Hilo. Also recover *CL* N and point P on the side roads. With the instrument set up on the mon-

**Figure 7-8** Connecting road traverse. *Courtesy of Ravi Thukaram and Kim Berry*

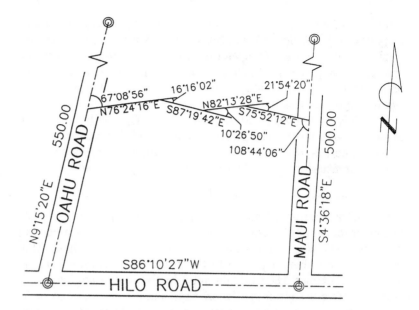

ument at Oahu, set a sight at point N. BE SURE TO PUT OUT TRAFFIC WARNING SIGNS AND WEAR FLUORESCENT VESTS WHENEVER WORKING IN A STREET. Set a mirror at approximately 550 ft. from the instrument on the *CL* of Oahu and measure the distance with the EDM. Now move the mirror the difference between the measurement and 550.00 ft. using the steel tape. Then remeasure the distance with the EDM and adjust the mirror the proper amount to be exactly 550.00. Check the line again to be sure the point is on *CL*. When distance and line are correct, drive a spike into the pavement directly under the mirror and punch the spike with a center point. Then move the instrument ahead and set up over the spike.

If the mirror is mounted on a tribrach on a tripod, the mirror can be detached using the tribrach release screw. The instrument can then be locked onto the tribrach without any setup error, as it will be in the same location as was the mirror.

Choose a point that can be seen from the instrument along the approximate line to be traversed and set up the mirror over this point. This is the point on traverse #1 (POT1) and the angle from the sight on point N on Oahu is turned to it. This is an angle to the right from N to #1 and should be recorded in the field book that way. Wind up the angle and record as 1, 2, 6, and mean in the book. Then read and record the distance to the mirror and the vertical angle to the center of the mirror. It is a good practice to read one vertical angle direct and another with the scope inverted, reading the distance both times as a check.

If the EDM has a foot/meter switch, it is a good check to read the distance in both feet and meters and record each in the book as such. Now, with POT1 set, move the instrument to the monument at Hilo and Maui and set the point at 500.00 along the *CL* of Maui. This could have been done previously but, this way, the instrument will not need to be set up on Oahu again. Also, the sight point N needs to be set (occupied) only once. Whenever a sight must be reset there is an opportunity for error.

Now set up the instrument on POT1 and take a sight on 0+00. The sight should be moved ahead, off the roadway, and reset on line given by the instrument. This eliminates the need to have anyone in the street any longer than necessary.

Set a second POT along the general line to Maui Road and set up the mirror over this POT. Sight the instrument on the POT1 with the scope inverted, then plunge the scope to look ahead down the line with the scope right side up. Release the upper motion and turn the scope to the right. This is called "deflected right." The motion is locked and the angle read.

Now release the lower motion and turn the instrument to point at the backsight. The scope should be direct and the first angle locked in the upper motion. Sight on the target, using the lower motion, and release the upper motion. Now transit (reverse) the scope and sight toward the foresight. Lock the upper motion and read the angle. Repeat for angles 3 through 6 and record the first, second, sixth, and mean in the field book. The distance and vertical angle can now be read to POT2 and the instrument moved ahead.

The angles should be recorded as a deflection angle to the right of 16-16-02. An angle to the right at this point would be 196-16-02, so you can see the difference it would make if you forget to record the type and direction of angle turned.

If you have three tripods, three tribrachs, and two mirrors, the instrument and mirrors can be leap-frogged ahead eliminating setup error. Leave the tripod and tribrach set up on POTs 1 and 2. Move the instrument ahead, lock it to the tribrach on POT2, and lock the mirror onto the tribrach on POT1. Set up another tripod and mirror on POT3, which should be set to be intervisible between POT2 and the point previously set in Maui Road. Sight the instrument on POT1 and turn a deflection angle to the left to POT3. Record the angle as a deflection angle to the left and measure and record the distance.

Move the instrument ahead to POT3 while the tripod from POT1 is moved ahead to the point in Maui Road. The mirror was placed on the tribrach on POT2 when it was passed, so move the tripod ahead to use with the mirror from POT3 to make the set up on Maui. Record the angle and distance and move the instrument to the tripod at the final point on the *CL* of Maui Road. Move the tripod and tribrach from POT2 to the monument at Hilo and Maui. Use the mirror from the setup point for the measurement. Turn an angle to the left to the monument from POT3. A deflection angle of 71-15-54 to the right can also be turned. Sometimes, as a check, turn both angles and then add to be sure they total 180°.

Because this is not a closed polygon a mathematical check of the angles cannot be done. Here, you must calculate the bearings of each course and compare them with the record closing bearing. The record bearing of Oahu Road is N 9-15-20 E. You turn an angle right of 67-08-56. Adding the two angles, you get a bearing from 0+00 to POT1 of N 76-24-16 E. At POT1, you turn a deflection angle right of 16-16-02. Adding that angle to the back bearing, you get 92-40-18. As the total angle is over 90°, this must be subtracted from 180°, which gives a bearing of S 87-19-42 E for POT1 to POT2. From POT2, you turn a deflection left of 10-26-50. Subtracting that from 92-40-18, you should get a bearing from POT2 to POT3 of N 82-13-28 E. At POT3, you turn a deflection right of 21-54-20 to the point on Maui Road. Adding 21-54-20 to N 82-13-28 E gives 104-07-48. Subtracting that from 180° gives a bearing of S 75-52-12 E from POT3 to Maui Road. Turning the angle to the monument at Hilo and Maui of 108-44-06 to the left is the same as a deflection to the right of 71-15-54 (180° minus 108-44-06). Subtracting 71-15-54 from S 75-52-12 E equals S 4-36-18 W. Congratulations! You closed flat again. Enjoy it while you can, from here it gets more complicated.

## TRAVERSE CALCULATIONS

Traverse calculations were formerly done only in the office by the person doing the computing. With the programmable calculator and lap-top computer came the capability to do the calculations in the field while the traverse is run. Knowing if a traverse closes before leaving the job site can be a real advantage. If the error in distance and the bearing of that error is known, it is possible to find the bad measurement and correct it in the field before returning to the office. Blunders usually occur along one of the sides of the traverse. If the closing error and one of the traverse legs have the same bearing, that is usually a good place to start looking for the problem. Angular errors should be caught before picking up the instrument if the mean angle is calculated on the spot.

An interior angle traverse is calculated by first finding if the interior angles add up to $n - 2$ (180). In Figure 7-5 the interior angles are: Pt. A, 90-00-00; Pt. B, 152-05-05; Pt. C, 100-32-10; Pt. D, 81-36-35; and Pt. E, 115-46-10. There are five angles: 5 minus 2 equals 3. Three times 180° equals 540°. Therefore, there should be 540° in the interior angles of the traverse. Add the interior angles and compare them with 540. Any error in closure may be adjusted out by one of several methods:

1.  The error may be divided equally between each of the angles in the traverse.
2.  If the error is small it may be put where it will do the least harm: between the shortest sides of the traverse. In Figure 7-5, that would be at Point A, if the error was only a few seconds. From A to B is the shortest leg of the traverse. An angle of 1 minute makes .1 of a foot in alignment at a distance of 300 feet. Therefore, an error of 15 seconds in 345 feet would only be about .03 feet.
3.  A weighted adjustment can be made and the error can be divided unequally between the points of the traverse. For instance, if the line between B and C was partially obstructed by vegetation, the angle at B might get more of the adjustment than the angle at A, which was unobstructed. If there were heat waves between D and E, the angles at D and E might get more of the adjustment than B.

After the interior angles have been adjusted, be sure to add them up again to check for errors. If you have twice as much error as you started with, you went the wrong way with the correction. Once all the error is adjusted out, the bearings of the sides of the traverse can be calculated.

## A BASIS OF BEARING

Every traverse must have a basis of bearing (BOB) to calculate the bearings of the individual sides. The BOB may be taken from an adjoining recorded bearing. The *CL* bearing of the fronting may be used. The record bearing of one of the courses in the legal description may be used in a lot survey. If no other bearing is available, an assumed bearing may be assigned to one of the sides. Care must be taken on assumed bearings that they are not mistaken for actual bearings of the lines. In some cases an astronomical bearing of the starting line may be taken. Always be sure to show the BOB on any plan and give its source.

The field notes in Figure 7-9 show the angles and distances measured in the field by the survey party. Take the traverse through, step-by-step, from field to finish. Notice that the party chief checked the horizontal angles for closure before leaving the job site. This can save a return trip to the site if the angles do not close within tolerance. The field notes are given to the person that will do the calculations in the office. That person will transfer the field data to a traverse calculation sheet such as Figure 7-10. The field angles are recorded in the proper column and added to check the closure.

**A**

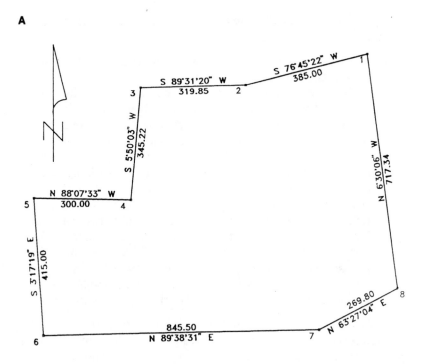

**Figure 7-9** A) Plans for an eight-sided traverse. *Courtesy of Ravi Thukaram and Kim Berry;* B) field notes for an eight-sided traverse

**B**

| Traverse 7-9 | | | | pg.1 of 3 |
|---|---|---|---|---|
| ⊼ @ "1" B.S. "8" F.S. "2" ∠ rt | | | | |
| | 1 - 23-15-15 | | | |
| | 2 - 46-30-30 | | | |
| | 6 - 139-31-36 | | | |
| | M - 23-15-16 | | | |
| dist. 1-2 385.62 @ 86-45 | | | | |
| ⊼ @ "2" B.S. "1" F.S. "3" ∠ rt | | | | |
| | 1 - 192-45-50 | | | |
| | 2 - 25-31-45 | | | |
| | 6 - 76-35-13 | | | |
| | M - 192-45-52 | | | |
| dist. 2-3 320.70 @ 94-10 | | | | |
| ⊼ @ "3" B.S. "2" F.S. "4" ∠ rt | | | | |
| | 1 - 96-18-35 | | | |
| | 2 - 192-37-10 | | | |
| | 6 - 217-51-36 | | | |
| | M - 96-18-36 | | | |

| Traverse 7-9 | | | pg. 2 of 3 |
|---|---|---|---|
| dist. 3-4 345.99 @ 86-10 | | | |
| ⊼ @ "4" B.S. "3" F.S. "5" ∠ rt. | | | |
| | 1 - 266-02-20 | | |
| | 2 - 172-04-35 | | |
| | 6 - 156-13-48 | | |
| | M - 266-02-18 | | |
| dist. 4-5 300.00 @ 90-00 | | | |
| ⊼ @ "5" B.S. "4" F.S. "6" ∠ rt. | | | |
| | 1 - 84-50-10 | | |
| | 2 - 169-40-15 | | |
| | 6 - 149-00-48 | | |
| | M - 84-50-08 | | |
| dist. 5-6 415.00 @ 90-00 | | | |
| ⊼ @ "6" B.S. "5" F.S. "7" ∠ rt. | | | |
| | 1 - 92-55-45 | | |
| | 2 - 185-51-25 | | |
| | 6 - 197-34-18 | | |
| | M - 92-55-43 | | |

| Traverse 7-9 | | | pg. 3 of 3 |
|---|---|---|---|
| dist. 6-7 845.59 @ 89-10 | | | |
| ⊼ @ "7" B.S. "6" F.S. "8" ∠ rt | | | |
| | 1 - 153-48-25 | | |
| | 2 - 307-36-50 | | |
| | 6 - 202-50-40 | | |
| | M - 153-48-27 | | |
| dist. 7-8 272.56 @ 98-10 | | | |
| ⊼ @ "8" B.S. "7" F.S. "1" ∠ rt. | | | |
| | 1 - 110-02-50 | | |
| | 2 - 220-05-40 | | |
| | 6 - 300-17-50 | | |
| | M - 110-02-58 | | |
| dist. 8-1 717.36 @ 89-38 | | | |
| | 1080 - 1079 -59-27 = 33" | | |
| | error = 33 seconds | | |
| | 2 - 185-51-25 | | |

**CLIENT:**  **JOB NO.:** 7-10
**LOCATION:**  **BK. NO.:**

| STATION | DISTANCE | BEARING | ∢ LT. | ∢ RT. | LAT. | DEPT. | Ng | Eg |
|---|---|---|---|---|---|---|---|---|
| 1 | | | | 83-15-25 | - 0.0024 / 88.1895 | +0.0111 / 374.7634 | | |
| 2 | 385.00 | S 76-45-29 W | | 192-45-52 | -0.0020 / -2.4927 | +0.0092 / 319.8403 | | |
| 3 | 319.85 | S 84-31-25 W | | 96-18-36 | -0.0022 / 343.4315 | +0.0100 / 35.0948 | | |
| 4 | 345.22 | S 5-50-05 W | | 266-02-18 | -0.0019 / +9.8114 | +0.0087 / 299.8395 | | |
| 5 | 300.00 | N 88-07-33 W | | 84-50-08 | -0.0026 / -414.3164 | +0.0120 / 23.8107 | | |
| 6 | 415.00 | S 3-17-21 E | | 92-55-43 | -0.0053 / +5.3042 | +0.0244 / 845.4834 | | |
| 7 | 845.50 | N 89-38-26 E | | 153-48-27 | -0.0017 / +120.5983 | +0.0578 / 241.3464 | | |
| 8 | 269.80 | N 63-26-57 E | | 110-02-58 | -0.0045 / +712.7388 | +0.0207 / 81.2063 | | |
| 1 | 717.35 | N 6-30-00 W | Total | 1079-59-27 | error | | | |
| | | | n-2 (180) Σ deg. | 1080-00-00 | +0.0225 | -0.1038 | | |
| | 3597.72 | Total | error | | correct | | | |
| | | | | ~ 33" | 0.0000063 | +0.0000289 | | |
| 1 | | | correct. | +4.125" x | 88.1920 | 374.7523 | 1000.0000 | 5000.0000 |
| 2 | | | | 83-15-29 | -2.4947 | 319.8311 | 911.8080 | 4625.2477 |
| 3 | | | | 192-45-56 | 343.4337 | 35.0848 | 909.3133 | 4305.4166 |
| 4 | | | | 96-18-40 | +9.8094 | 299.8308 | 565.8796 | 4270.3318 |
| 5 | | | | 266-02-22 | -414.3190 | 23.8227 | 575.6891 | 3970.5010 |
| 6 | | | | 84-50-12 | +5.2989 | 845.5078 | 161.3701 | 3994.3237 |
| 7 | | | | 92-55-47 | +120.5966 | 241.3542 | 166.6690 | 4839.8315 |
| 8 | | | | 153-48-31 | +712.7343 | 81.1856 | 287.2656 | 5081.1857 |
| 1 | | | | 110-03-03 | | | 100000 | 5000.00 |
| | | | | 1080-00-00 | -0.0001 | +0.0001 | | |

**Figure 7-10** Coordinate worksheet

In the traverse shown, the angles do not close by 33 seconds. As the error is in a negative direction, a correction of 4.125 seconds needs to be added to each of the raw angles to get a total of 1080°. There are eight angles, so 4 seconds are added to each angle and an extra second must be added to every eighth angle to come out even

$$(4 \times 8 = 32 + 1 = 33 \text{ sec.}).$$

Next, the slope distances must be reduced to horizontal and recorded in the distance column. The BOB from point 8 to point 1 is

**LATITUDES AND DEPARTURES**

**Figure 7-11** Latitude and departure sine and cosine

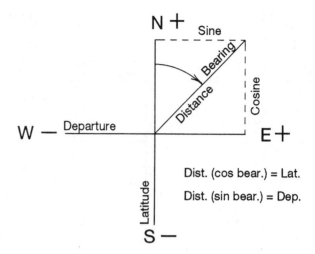

recorded in the bearing column between 8 and 1. The bearings can then be calculated and recorded in the spaces between stations. Notice that between each station, there is room for a distance and bearing and across the page, the latitudes and departures also line up. These are the data between points, whereas at the point, there is a space for angle and coordinates of the point only. Next, the distance is multiplied times the cosine of the bearing to get the latitude and is recorded in the latitude column. The distance is next multiplied times the sine of the bearing to get the departure, which is recorded in the departure column. Each of the latitudes will have a sign. The sign will be positive for north and negative for south. The departures will also have signs; positive for east and negative for west (see Figure 7-11). Next, add the latitudes algebraically, paying attention to the sign of the result. Now add the departures algebraically, again recording the sign of the sum.

**BALANCING OR ADJUSTING THE TRAVERSE**

There are four common methods of balancing or adjusting the traverse to minimize the effect of error. They are (1) the compass rule, (2) the transit rule, (3) the Crandall rule, and (4) the least squares adjustment. The American Congress on Surveying and Mapping defines balancing a survey as, "Distributing corrections through a traverse to eliminate the errors of closure." Note that the balance, or adjustment, does not remove the error but only distributes it.

The **compass (or Bowditch) rule** assumes that the error is caused equally by angle and distance. It is the most commonly used balancing method.

The **transit rule** assumes the error is greater in the measurement of distance than in the measurement of angle. This is not always the case, so the balance works best on surveys that run north-south or east-west.

The **Crandall rule** makes the same assumption as the transit rule but uses a modified **least squares adjustment** to give better results. Many computer programs have a Crandall adjustment and the compass and transit balance in their traverse adjustment menu.

---

When you count the number of sides in a traverse, the count is a certain number. There are four sides in a square, no more, no less. When you measure those sides, no matter how carefully, an amount of uncertainty is introduced. Accuracy is the ability to arrive at the *true* distance or number. **Precision** is how close to each other the measurements are made. Precision indicates that you are consistent in your work. If your equipment is out of adjustment, you may be consistent (precise) but you won't be accurate. If you take a large number of precise measurements, some of these measurements will probably be very close to the true, or accurate, distance. If the measured distances are added together and divided by the number of distances, a mean distance is found. This mean distance will be close to the true distance. When the measured distances are subtracted from the mean distance, a residual is found that can be used to determine the standard error of the group of measurements. If the measurements are given weights, a higher number to the more precise, a lower number to those that are less precise, and the rules of statistics applied, a number very close to the true distance can be calculated.

The **least squares method of adjustment** uses the squares of the weighted residuals to arrive at the most probable measurement. There are two types of least squares adjustments: adjustment of observations and adjustment of indirect observations. Both use the principle *the sum of the squares of the observational residuals must be a minimum.*

The easiest mathematical solution for least squares is through the use of matrix algebra. The solution is a time-consuming series of calculations. However, there are now several good computer programs on the market that will do a least squares adjustment in only a few minutes. By using these programs, the preplanning of field observations for meeting the requirements of positional tolerance is possible. The least squares method has the advantage of using all the field data—distances, angles, and reference azimuths—and solving for all the unknown data simultaneously. All the other methods separate the angles and adjust them first and then consider the distances.

For this traverse, the compass rule will be used to balance the latitudes and departures. The compass rule states that the correc-

**ACCURACY AND PRECISION**

tion to be applied to the latitude or departure of each course is to the total error in latitude or departure as the length of the corresponding traverse side is to the total of all the sides of the traverse. When all the latitudes in the latitude column are added together using proper signs, you get a residual error of +0.0225. If you divide this error by the total distance surveyed of 3597.72, you get 0.0000063. This is the correction per foot to the distance. As your error had a positive direction, you must give the correction the opposite sign to remove the error. If you give the wrong sign to the correction, you will end up with twice the error when the result is checked. If you multiply −0.0000063 by the distance from 1 to 2 of 385.00, you get a correction of −0.0024 to be applied to the latitude of 1 to 2 of −88.1895; adding −0.0024 to −88.1895 gives a corrected latitude for 1 to 2 of −88.1920. Check to make certain that the corrected latitude is smaller than the uncorrected latitude because you must increase the north latitudes and decrease south latitudes. Continue through the other seven latitudes, recording the corrected latitudes at the bottom of the latitude column.

After correcting all the latitudes, be sure to add up the new latitudes to see if they total zero. If you do not carry the calculations to enough decimal places, there may be a small residual error. If the remaining error is at least two zeros more than your least measurement, you can probably ignore it.

Now, do the same with the departures. When the departures are added up, there is an error of −0.1038. When you divide −0.1038 by 3597.72, you get a correction of +0.0000289. If you multiply +0.0000289 by 385.00, you get a correction of +0.0111 for the departure. Adding +0.0111 to −374.7634, you get −374.7523 for the corrected departure from 1 to 2. Continue through the other seven departures in the same way, and be sure to add them up when they have all been corrected.

Do not be concerned if this seems complicated. Most calculations are now made on a computer by a menu-driven program. The raw field data is either entered by hand or, if the total station has a data collector, it may be downloaded directly to the computer through an interface. Hand-held calculators are used in the field to do the calculations and sometimes to record data. The increasing use of the laptop computer has made solving difficult computations in the field commonplace.

## CALCULATION OF THE COORDINATES

Now that you have corrected latitudes and departures, the next step is the calculation of the coordinates of the stations. On the right side of the traverse computation sheet are two columns: Ng and Eg (see Figure 7-10). These stand for **northing** and **easting** or ordinate and abscissa or Y and X coordinates. If you have had algebra, you will recognize the **rectangular coordinate system.** If you have not, this might be a good time to start.

If coordinates are not given for the point of beginning, they may be assigned. Assigning coordinates to a point creates a **local coordinate system.** The coordinates assigned should be great enough to prevent getting into negative coordinates on the opposite side of the traverse. It is wise to assign a different starting coordinate to the Eg so that you will be able to recognize the Ngs from the Egs in the computer printout just by their number. To calculate the coordinates for each station or traverse point, add the corrected latitude or departure to the previous Ng or Eg. In Figure 7-10, the assigned coordinates of station 1 are 1000.0000 Ng and 5000.0000 Eg. Adding the latitude of station 1 to station 2 of −88.1020 to the north coordinate of station 1 gives a Ng for station 2 of 911.8080. Notice the calculations are carried out to four decimal places. This allows rounding off to three decimal places or one more than the number of places used in the measurements. Adding the departure of station 1 to 2 to the Eg of station 1 of 5000.0000 gives an Eg for station 2 of 4625.2477. Continue through the other stations and be sure to close back to the starting point of 1000–5000 to check for any error in calculations.

By now, you are probably wondering just what good all this calculating does. What can be done with the information that has been generated? If the BOB used was a record bearing that must be held, then it must be checked to see if it has been rotated by the adjustment.

Use the calculated coordinates to find the new distances and bearings by inversing between the coordinates of the adjoining points. Inverse calculations are just what they sound like. The corrected latitudes and departures were used to get the coordinates, and coordinates can be used to get latitudes and departures between points.

**INVERSE CALCULATIONS**

Because the latitudes and departures have already been calculated, they can be used in the inverse calculations. If you only have the coordinates, you have to subtract Ng 1 from Ng 2 to obtain the latitude from 1 to 2. Remember, north is plus and south is minus. In a like manner, subtract Eg 1 from Eg 2 to get the departure from 1 to 2, again paying attention to the sign—east is plus and west is minus. If you have had trigonometry, you know that the side opposite divided by the side adjacent equals the tangent of the angle. The departure is the side opposite the bearing angle, so if you divide the departure by the latitude or side adjacent to the bearing angle, you get the tangent of the bearing angle.

$$374.7523 / 88.1920 = 4.2493:$$
$$\arctan 4.2493 = 76.7573° = 76\text{-}45\text{-}26$$

Because both latitude and departure had minus signs, the angle is in the southwest quadrant or S 76-45-26 W.

Find the distance between the two points by using either trigonometry functions to solve for the hypotenuse or the Pythagorean theorem. The trigonometry function to use depends on the bearing angle. Try to use the function that will use the longest side. If the latitude is longest, use the latitude divided by the cosine of the bearing angle. Because the legs are nearly of equal length, this is not necessary but for angles under 15° or over 85°, the strength of figure becomes very weak. The Pythagorean theorem ($a^2 + b^2 = c^2$) is easy to use with a pocket calculator. Enter the latitude and push $x^2$, push $+$ and enter the departure, and push $x^2$ again. Now push $=$ and then $\sqrt{x}$. The answer is the distance between the stations. For RPN calculators, enter the latitude and push $x^2$, enter the departure and push $x^2$, now push $+$ and $\sqrt{x}$. The distance is displayed on the calculator.

If you inverse between station 8 and station 1, you find that the bearing from 8 to 1 has rotated from N 6-30-00 W to N 6-29-54 W or 8-sec. clockwise. To get back to N 6-30-00 W, you must rotate the adjusted traverse by 8-sec. counterclockwise. You cannot just change the bearings without calculating new rotated coordinates if you wish to use the coordinates for other calculations. If all you need are the bearings and distances, the bearings can be changed by 8 sec. and you will be done. Be sure to apply the correction in the proper direction and check for closure. If you are calculating new coordinates, be sure to inverse and check the rotated bearings and distance against the originals.

If the error of closure is greater than the allowable limits, it is sometimes possible to isolate the bad leg by cutting across the traverse and creating two or more traverses that are smaller. These smaller traverses can each be closed on themselves and, if no error is found, the points in the main traverse that are included in the subtraverse can be considered good. When one of the subtraverses fails to close, check the points on the main traverse that are included in it. The blunder will usually be in one of them. This method of error trapping can be used to find a blunder, but it will not help if the error was cumulative because of faulty survey methods or equipment.

## THE "FLY POINT"

On occasion, one of the points you need in your traverse cannot be occupied. A traverse point can be set close enough to the point so that angle and distance can be measured in one setup. The angle

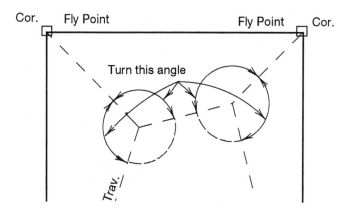

**Figure 7-12** Fly point or eccentric to a corner

should be measured from the backsight to the **fly point** and also from the fly point to the foresight. These angles and the traverse angle should add up to 360° (see Figure 7-12).

Distance should be checked at least twice to be sure it is correct. A fly point is not a part of the main traverse and the traverse closure will not show a mistake in its calculation.

## THE AREA OF A POLYGON

After the corners of a closed traverse have been balanced, the area can be found. There are several ways to find the area of a polygon using graphical or mathematical methods. Graphically, a scale drawing can be made on a grid paper and the number of squares counted. The number of square feet in each square is multiplied by the number of squares and partial squares and the result is the approximate area in square feet.

## POLAR PLANIMETER

You can also use a polar planimeter to find the number of square inches in a scale drawing. The scale size is multiplied to find the number of square feet in 1 square inch. The result is then multiplied by the number of square inches to find the area of the figure. On a 50 (1 in. = 50 ft.) scale drawing there would be 50 times 50 or 2500 square feet to a square inch. The closed figure can also be broken down into a series of rectangles and triangles and the areas of those calculated and added up to get a total area.

## DOUBLE MERIDIAN DISTANCE

The **double meridian distance** (DMD) method of finding area is commonly used and appears on many survey tests. The DMD uses the latitudes and departures of the adjusted traverse to calculate the area by using the area of a trapezoid. The area of a trapezoid is one half the sum of the bases times the altitude. In the DMD,

**Figure 7-13** Departure and DMD distance

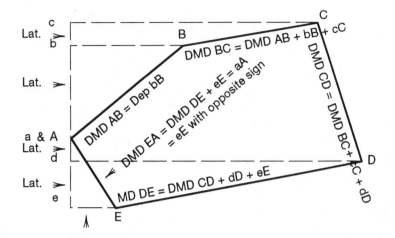

the median of the trapezoid is found using the departures with the proper algebraic sign. Positive for east, negative for west from the midpoint of the course to a reference meridian running north and south through the westernmost point of the traverse (see Figure 7-13).

You can find the area by multiplying the DMD of each course by the latitude of that course. North latitudes are positive and south latitudes are negative. Add the areas algebraically and divide by 2 to get the total area enclosed by the traverse. To calculate the DMD of each course:

- The DMD of the first course is equal to the departure of the first course.
- The DMD of each succeeding course is equal to the DMD of the preceding course plus the departure of the preceding course plus the departure of the course itself.
- The DMD of the last course is the departure of the last course with the sign changed (west would be positive and east would be negative).

## AREA BY USING COORDINATES

Area by using coordinates is widely used now that the hand-held calculator is in common use. The formula for a four-sided traverse would be:

$$\text{Area } 1,2,3,4, = \frac{1}{2} \ (x2y1 - x1y2 + x3y2 - x2y3 + x4y3 - x3y4 + x1y4 - x4y1).$$

If this overloads your circuits, then arrange the coordinates as on the calculation sheet (see Figure 7-10). Now multiply N1 by E2 and add that result to N2 times E3, and so on through the list. Be sure to multiply the last Ng to the first Eg (N (last) times E1). Repeat with the other column, E1 times N2 plus E2 times N3. Again be sure to include the last Eg times the first Ng (E (last) times N1). Subtract the first total from the second total without regard to sign. You cannot have a negative area. Now divide the result by 2. The answer is the area in square feet. If you need the area in acres, divide the square feet by 43,560 to get acres. There are 43,560 square feet in one acre. (Note: The keystrokes for a RPN calculator are: N1, enter, E2, ×, N2, enter, E3, ×, +, N3, enter, E4, ×, +, N4, enter, E5, ×, +, etc. The totals move up in the stack and will drop down when the last plus is entered. The total of the first column is now in the Y register and the last column in the X register. Push − to subtract 1 and 2, ÷ to get the area in square feet.) Most computer programs give the distance traversed and total area as part of the traverse closure program.

## Review Questions

1. State the difference between an open and a closed traverse.
2. What is a connecting traverse?
3. When winding up an angle six times, which angles should be recorded in the field book?
4. When setting a sight for line, what should you check before moving ahead?
5. Why is using a short backsight distance undesirable?
6. How many degrees should the interior angles of a six-sided traverse add up to?
7. What is a POB?
8. How far is it between Sta. 2 + 46.20 and Sta. 13 + 96.02?
9. What equipment is necessary for safety when working in a street?
10. What should you check for before picking up a theodolite or mirror mounted on a tribrach?
11. What is a POT?
12. What is a deflection angle?
13. The distance between points is multiplied by the cosine of the bearing angle to get the _____ of the course and by the sine of the bearing angle to get the _____.
14. What are the four common methods of balancing, or adjusting, a traverse?
15. On a counterclockwise traverse, the following mean angles were turned to the right. The distances were doubled, measured with an EDM, and reduced to horizontal distances. The basis of bearing is the line between points 1 and 2 and is north. The assumed coordinates for point 1 are north 1000.0000 and east 5000.0000. Calculate the closure and adjust the traverse using the compass rule. Calculate coordinates for the corners, find the

bearings, and adjust distances for each course. Hold 1 to 2 as the bearing north and translate and rotate the adjusted coordinates to match. Note: BS = backsight, FS = foresight.

Inst. at point 1 BS 12 FS 2 angle right
     66°-42'-10" dist. 582.20

Inst. @ #2 BS 1 FS 3 ang. rt
     180°-00'-30" dist. 460.00

Inst. @ #3 BS 2 FS 4 ang. rt
     179°-58'-30" dist. 30.00

Inst. @ #4 BS 3 FS 5 ang. rt
     90°-29'-00" dist. 460.65

Inst. @ #5 BS 4 FS 6 ang. rt
     203°-38'-30" dist. 204.11

Inst. @ #6 BS 5 FS 7 ang. rt
     107°-13'-40" dist. 318.80

Inst. @ #7 BS 6 FS 8 ang. rt
     260°-48'-00" dist. 290.04

Inst. @ #8 BS 7 FS 9 ang. rt
     78°-42'-00" dist. 285.30

Inst. @ #9 BS 8 FS 10 ang. rt
     180°-00'-00" dist. 265.74

Inst. @ #10 BS 9 FS 11 ang. rt
     92°-29'-00" dist. 419.60

Inst. @ #11 BS 10 FS 12 ang. rt
     179°-57'-40" dist. 715.56

Inst. @ #12 BS 11 FS 1 ang. rt
     180°-01'-10" dist. 279.10

# Topography – Field and Office

**8**

## Objectives

After completing this chapter and the field practice, the student should be able to:

1. Measure a distance using the stadia hairs in a transit/theodolite and a stadia rod and reduce the data to horizontal and vertical distances.
2. Gather data for a transit topo using a transit/theodolite and rod or a total station and prism pole.
3. Plot the data gathered using a protractor and scale.
4. Plot the data gathered using a calculator and scale.
5. Draft a topographical map using the plotted points.
6. Set premarks for aerial photogrammetry.
7. Draft a plan and profile map using the proper drafting paper and scale.
8. Specify the information to be gathered in the field to produce an A.L.T.A. plat.
9. Produce a finished topographical map using the proper size and type of lines to differentiate between the contour and elevations.

---

Previous chapters covered how to locate points vertically and horizontally. A three-dimensional picture of the terrain can be formed by combining the horizontal and vertical position for one point with those of nearby points. When the points that lie on the same elevation are joined by a line, the **contours** of the land form

**A THREE-DIMENSIONAL PICTURE**

what can be seen as hills and valleys. The closer together the contour lines are, the steeper the terrain. With practice, you will be able to read a **topographic map** as if it were a three-dimensional model.

## AN "INSTRUMENT TOPOGRAPHICALL"

An "instrument topographicall" was first described by Digges in 1575. William Leybourn, in *The Compleat Surveyor,* published in London in 1653, describes a "playne table" used for making topographical maps. The plane table used in the sixteenth century is nearly identical to the one in use today.

Although the use of the plane table has dwindled in the past decade because of the increased use of topo methods with a total station, it could be useful for small, crowded sites. The conventional stadia alidade is being replaced by the self-reducing alidade and the alidade-mounted EDM with automatic slope reduction. The major advantage of using a plane table for topo is the ability to find gaps in the map while you are still set up in the field. If you are doing a topo of a site with many trees to locate, the rod can be placed alongside the tree trunk and the distance measured. The location is then taken to the center of the tree with the alidade. The size and type of tree can then be noted next to the location on the topo, while the rod carrier moves to the next tree. The major disadvantage of the plane table is a stiff back that results from leaning over the table all day long.

## STADIA OR AN EDM

In an area that gets a good deal of rain (such as Washington or Oregon), the plane table presents another disadvantage: the paper gets wet and smears. This can be overcome by using drafting film instead of paper, but the alidade sliding on the wet surface may still cause problems with smearing. Under wet conditions, a transit or theodolite with stadia or an EDM may be used to take the shots and the information recorded in a "Rite-in-the-Rain" field book. In heavy rain, put the field book in a plastic bag to keep it from getting so wet that the pencil tears the surface. You can see through the plastic to keep notes. A plastic bag placed over the instrument, with a hole left for the lens, and fastened with rubber bands will help keep the water off. The page heading must have several columns to record. The stadia interval and distance, the horizontal angle from the rod to the backsight, and the vertical angle (plus, minus, or level) is required. Also required are the rod reading (shooting the HI will simplify the calculations) and the vertical differences from the instrument or station height to the center

**Figure 8-1** Transit topo field notes

| | TRANSIT | TOPO | | | |
|---|---|---|---|---|---|
| SI/HD | Hor ∠ | Vert ∠ | Rod | Vt. Dif | Elev |
| | △ @ Sta. A  Rt. ∠ MU 5.35 | | | | |
| | 0-00 | 0-00 | +11.25 | | 540.56 |
| 1.52 | 25-34 | -4-20 | - | | |
| 1.93 | 28-40 | -3-15 | - | | |
| 2.36 | 30-10 | - | 10.4 | | |
| 2.80 | 32-25 | - | 2.4 | | |
| 3.30 | 35-48 | +2-35 | - | | |
| 3.75 | 38-10 | +3-10 | - | | |
| 4.15 | 40-00 | +3-40 | - | | |
| 4.65 | 42-30 | +4-08 | - | | |
| | | | | | |

hair reading. The elevation of the point must also be shown (see Figure 8-1). Slope distances must be reduced to horizontal before being **plotted**. To avoid confusion, the term "measure up" (MU) is now used to record the distance from the setup point to the center of the telescope. Previously, this distance was shown as HI as used in a level run to indicate the elevation of the center of the telescope.

To reduce stadia distances to horizontal, let

SI = stadia interval, (upper wire reading minus lower wire reading)

HD = horizontal distance from the instrument to the rod, and

VD = vertical distance from the center of vertical circle on the instrument to the rod reading on the center hair. Let

VA = the vertical angle, + or −, to the center hair from the HI (or from the station elevation if the center hair was set on the HI)

HD = 100(SI)cos²VA,

VD = 100SI(cosVA) (sinVA)

MU = measure up.

Stadia reduction tables can also be used (see table in Figure 8-2).

If the distances were shot with a total station to a prism on an adjustable rod set at the HI of the instrument, just multiply the slope distance by the sine of the vertical angle. This will give you the vertical difference. Multiply by the cosine to get the horizontal distance. If the HI was not sighted with the center hair, the rod reading must be either added to or subtracted from the vertical difference (see Figure 8-3).

## STADIA COEFFICIENTS, VERTICAL ROD.

Cos²v and ½ sin 2v.

| ′ | 4° Hor. dist. | 4° Diff. elev. | 5° Hor. dist. | 5° Diff. elev. | 6° Hor. dist. | 6° Diff. elev. | 7° Hor. dist. | 7° Diff. elev. | 8° Hor. dist. | 8° Diff. elev. | 9° Hor. dist. | 9° Diff. elev. | 10° Hor. dist. | 10° Diff. elev. | 11° Hor. dist. | 11° Diff. elev. |
|---|---|---|---|---|---|---|---|---|---|---|---|---|---|---|---|---|
| 0 | .9951 | .0696 | .9924 | .0868 | .9891 | .1040 | .9851 | .1210 | .9806 | .1378 | .9755 | .1545 | .9698 | .1710 | .9636 | .1873 |
| 2 | .9951 | .0702 | .9923 | .0874 | .9890 | .1045 | .9850 | .1215 | .9805 | .1384 | .9753 | .1551 | .9696 | .1716 | .9634 | .1878 |
| 4 | .9950 | .0707 | .9922 | .0880 | .9888 | .1051 | .9848 | .1221 | .9803 | .1389 | .9752 | .1556 | .9694 | .1721 | .9632 | .1884 |
| 6 | .9949 | .0713 | .9921 | .0885 | .9887 | .1057 | .9847 | .1226 | .9801 | .1395 | .9750 | .1562 | .9692 | .1726 | .9629 | .1889 |
| 8 | .9948 | .0719 | .9920 | .0891 | .9886 | .1062 | .9846 | .1232 | .9800 | .1401 | .9748 | .1567 | .9690 | .1732 | .9627 | .1895 |
| 10 | .9947 | .0725 | .9919 | .0897 | .9885 | .1068 | .9844 | .1238 | .9798 | .1406 | .9746 | .1573 | .9688 | .1737 | .9625 | .1900 |
| 12 | .9946 | .0730 | .9918 | .0903 | .9883 | .1074 | .9843 | .1243 | .9797 | .1412 | .9744 | .1578 | .9686 | .1743 | .9623 | .1905 |
| 14 | .9946 | .0736 | .9917 | .0908 | .9882 | .1079 | .9841 | .1249 | .9795 | .1417 | .9743 | .1584 | .9684 | .1748 | .9621 | .1911 |
| 16 | .9945 | .0742 | .9916 | .0914 | .9881 | .1085 | .9840 | .1255 | .9793 | .1423 | .9741 | .1589 | .9682 | .1754 | .9618 | .1916 |
| 18 | .9944 | .0748 | .9915 | .0920 | .9880 | .1091 | .9839 | .1260 | .9792 | .1428 | .9739 | .1595 | .9680 | .1759 | .9616 | .1921 |
| 20 | .9943 | .0753 | .9914 | .0925 | .9878 | .1096 | .9837 | .1266 | .9790 | .1434 | .9737 | .1600 | .9678 | .1765 | .9614 | .1927 |
| 22 | .9942 | .0759 | .9913 | .0931 | .9877 | .1102 | .9836 | .1272 | .9788 | .1440 | .9735 | .1606 | .9676 | .1770 | .9612 | .1932 |
| 24 | .9941 | .0765 | .9911 | .0937 | .9876 | .1108 | .9834 | .1277 | .9787 | .1445 | .9733 | .1611 | .9674 | .1776 | .9609 | .1938 |
| 26 | .9940 | .0771 | .9910 | .0943 | .9874 | .1113 | .9833 | .1283 | .9785 | .1451 | .9731 | .1617 | .9672 | .1781 | .9607 | .1943 |
| 28 | .9939 | .0776 | .9909 | .0948 | .9873 | .1119 | .9831 | .1288 | .9783 | .1456 | .9729 | .1622 | .9670 | .1786 | .9605 | .1948 |
| 30 | .9938 | .0782 | .9908 | .0954 | .9872 | .1125 | .9829 | .1294 | .9782 | .1462 | .9728 | .1628 | .9668 | .1792 | .9603 | .1954 |
| 32 | .9938 | .0788 | .9907 | .0960 | .9871 | .1130 | .9828 | .1300 | .9780 | .1467 | .9726 | .1633 | .9666 | .1797 | .9600 | .1959 |
| 34 | .9937 | .0794 | .9906 | .0965 | .9869 | .1136 | .9827 | .1305 | .9778 | .1473 | .9724 | .1639 | .9664 | .1803 | .9598 | .1964 |
| 36 | .9936 | .0790 | .9905 | .0971 | .9868 | .1142 | .9825 | .1311 | .9776 | .1479 | .9722 | .1644 | .9662 | .1808 | .9596 | .1970 |
| 38 | .9935 | .0805 | .9904 | .0977 | .9867 | .1147 | .9824 | .1317 | .9775 | .1484 | .9720 | .1650 | .9660 | .1814 | .9593 | .1975 |
| 40 | .9934 | .0811 | .9903 | .0983 | .9865 | .1153 | .9822 | .1322 | .9773 | .1490 | .9718 | .1655 | .9657 | .1819 | .9591 | .1980 |
| 42 | .9933 | .0817 | .9901 | .0988 | .9864 | .1159 | .9820 | .1328 | .9771 | .1495 | .9716 | .1661 | .9655 | .1824 | .9589 | .1986 |
| 44 | .9932 | .0822 | .9900 | .0994 | .9863 | .1164 | .9819 | .1333 | .9769 | .1501 | .9714 | .1663 | .9653 | .1830 | .9586 | .1991 |
| 46 | .9931 | .0828 | .9899 | .1000 | .9861 | .1170 | .9817 | .1339 | .9768 | .1506 | .9712 | .1672 | .9651 | .1835 | .9584 | .1996 |
| 48 | .9930 | .0834 | .9898 | .1005 | .9860 | .1176 | .9816 | .1345 | .9766 | .1512 | .9710 | .1677 | .9649 | .1841 | .9582 | .2002 |
| 50 | .9929 | .0840 | .9897 | .1011 | .9858 | .1181 | .9814 | .1350 | .9764 | .1517 | .9708 | .1683 | .9647 | .1846 | .9579 | .2007 |
| 52 | .9928 | .0845 | .9896 | .1017 | .9857 | .1187 | .9813 | .1356 | .9762 | .1523 | .9706 | .1688 | .9645 | .1851 | .9577 | .2012 |
| 54 | .9927 | .0851 | .9894 | .1022 | .9856 | .1193 | .9811 | .1361 | .9761 | .1528 | .9704 | .1694 | .9642 | .1857 | .9575 | .2018 |
| 56 | .9926 | .0857 | .9893 | .1028 | .9854 | .1198 | .9810 | .1367 | .9759 | .1534 | .9702 | .1699 | .9640 | .1862 | .9572 | .2023 |
| 58 | .9925 | .0863 | .9892 | .1034 | .9853 | .1204 | .9808 | .1373 | .9757 | .1540 | .9700 | .1705 | .9638 | .1868 | .9570 | .2028 |
| 60 | .9924 | .0868 | .9891 | .1040 | .9851 | .1210 | .9806 | .1378 | .9755 | .1545 | .9698 | .1710 | .9636 | .1873 | .9568 | .2034 |
| c+f 0.75 | 0.75 | 0.06 | 0.75 | 0.07 | 0.75 | 0.08 | 0.74 | 0.10 | 0.74 | 0.11 | 0.74 | 0.12 | 0.74 | 0.14 | 0.73 | 0.15 |
| 1.00 | 1.00 | 0.08 | 0.99 | 0.09 | 0.99 | 0.11 | 0.99 | 0.13 | 0.99 | 0.15 | 0.99 | 0.16 | 0.98 | 0.18 | 0.98 | 0.20 |
| 1.25 | 1.25 | 0.10 | 1.24 | 0.11 | 1.24 | 0.14 | 1.24 | 0.16 | 1.23 | 0.18 | 1.23 | 0.21 | 1.23 | 0.23 | 1.22 | 0.25 |

$(c+f) \cos v$ and $(c+f) \sin v$.

Natural functions.

$(c+f) \cos v$ and $(c+f) \sin v$.

Natural functions.

**Figure 8-2** Stadia reduction tables. *Courtesy of U.S. Department of the Interior Land Management Bureau*

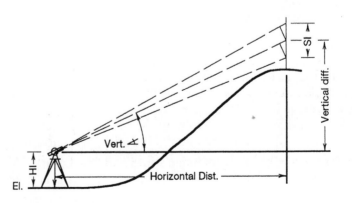

**Figure 8-3** Vertical angle stadia shot

The experience and skill of the rodperson will to a great degree determine the accuracy and speed of the topo. The instrument operator should try to anticipate the location of the next shot and be ready to read the rod as soon as it faces the instrument. If the rodperson is waved ahead as soon as the necessary information is recorded, the topo will proceed at a fast pace.

You, as the rodperson, should be planning the next shot while the instrument person is reading the rod. As soon as the instrument operator waves both arms overhead in a signal to move ahead, you move swiftly to the next point. Place the bottom of the rod on the point and allow your progress forward to raise the rod to a vertical position. Turn the rod face away from the instrument to prevent it being read until you are ready to take a shot. This allows the instrument operator time to align the scope on the rod. When the rod is where you want to take the shot from, face it so that it can be seen from the instrument position and raise one arm overhead. This signals the instrument person to read the rod.

The rod may be obstructed by foliage or some other object not visible to the rodperson. The instrument operator will signal the rodperson to move the rod to one side or the other by waving in the direction desired. When the rod is visible in the field of view, the instrument person signals and the rod is again faced toward the instrument. If the foot mark, or prism, is still obstructed, the instrument operator may signal to raise the rod, or prism. Raise one arm overhead repeatedly until the foot mark, or prism, can be seen. This is called "raise for red" because the full foot marks are usually printed in red ink or paint.

If the best line of sight is over the top of the extended rod, the instrument operators may signal to raise the rod an even number of feet by touching themselves on the foot with one hand. This is called a **boot** and should be carefully measured by the rodperson and the amount signaled to the instrument operator (see Figure 8-4). Sometimes an auxiliary position may be measured by the rodperson from the last shot. The distance toward or away from the instrument and the difference in elevation is signaled to the instrument operator.

**THE RODPERSON**

**Figure 8-4** Booting the rod

**WHERE TO TAKE THE SHOTS**

Experience and judgment on the part of the rodperson are required when deciding where to take the shots. If too many shots are taken, the plotting and field times are increased. If too few shots are taken, the topography may not be accurate. Usually, you should take a shot at each change in **grade** or grade break. Take a shot at the top of a slope and the bottom, or **toe**, of the slope (see Figure 8-5). Show top of bank, width and flow line of streams, whether running or dry. Show all man-made improvements, called

**Figure 8-5** Description of natural features

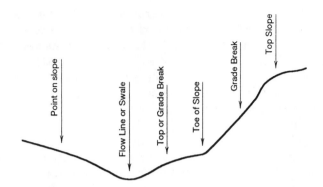

"culture," and edges of vegetation changes. Show trees individually that are above a certain size. State their species and diameter at breast height (dbh, 4.5-ft above ground). The high or low point of closed contour features such as hills or swales (dips in the ground similar to a pond but without water) should be shown.

Show the center line on streets, ¼ points, flow lines, the tops of curbs, and back of walk. On manholes, take a shot on the uphill and downhill rim and give the type: sanitary sewer, storm drain, electric vault, or cable. Open the manhole lid and record the size and direction of inlet and outlet pipes. Measure down from the rim to the flow line of each pipe (**invert**) with the rod and record the distance to the nearest hundredth of a foot. Show all shots on concrete to the hundredth and ground shots to the nearest tenth.

Give the location of utility poles and their type. BE VERY CAREFUL AROUND POWER POLES TO KEEP THE ROD AWAY FROM THE WIRES. Electricity can jump a very large gap to get to a ground such as a metal-faced rod or a wet one. Contact your local power company to see if they give classes on safety around power lines. Attend such a class or request that one be given at your school or company. Most utility poles have a number stamped on a metal strip. Be sure to record the number in your notes. Some poles have guy wires running to an anchor (dead man) set in the ground. Be sure to locate these.

Water valves or gas valves under metal lids can help to call the drafter's attention to underground utilities. Give the type and location of any you find. On storm drain catch basins, give the size of the structure and any outlet pipes you can find.

Locate at least two corners of buildings and make a drawing showing the shape and length of the exterior walls. Give the type of structure and number of stories. Angle points in fence lines will locate the fence. Be sure to give the height and type: wood, wire, chain link, barbed wire, etc. Old fence lines are sometimes good indicators of lines of occupation so be sure to show old fence posts, even if the fence is no longer standing. Show any signs of well-traveled roads or footpaths across the property. These may suggest an **easement** or **evidence** of a **prescriptive right** of use. Remember that you are the licensed surveyor's eyes in the field so be observant and report anything you feel may be important. It will take much practice running rod to become proficient at it, but in time it will become quite natural and your speed will also increase.

A grid may be appropriate on topos with little relief that is rather free of trees. On a **grid topo,** the area is divided into some convenient grid by using lath and flagging to mark off the distance. Usually a 100-ft. grid is sufficient to show the relief, with supplemental shots on any grade break between grid points. These supplemental points may be paced in or measured with the rod or pocket tape. Do not forget to watch for grade breaks between the rows as you are moving ahead along a row of lath. These can be located as right or left of the row you are working on. Sometimes it is possible to set lath further apart and use a double right angle prism to line in between points. Two laths in a row make it easy to line in on unmarked points.

**A GRID TOPO**

On topos that are steep but somewhat clear, a method called **trace contour** can be used. It decreases the amount of climbing up and down slopes by the rod carrier. With the trace contour method, you set a target on the rod at a foot mark that will give an even foot on the elevation of the contour line. The rod carrier then travels back and fourth across the slope instead of up and down. The target is moved to a different agreed-upon foot for each trip across. The elevation of the contour can be obtained at the end of each contour by setting the target with the instrument or by using the hand level. The elevation is then carried across the slope by the rod carrier, using a hand level to carry the same setting from point to point. Because all the shots are on the same elevation across the slope only, distance and horizontal angles are needed. With the use of the slope-reducing total stations, this method can be very fast. It has the advantage of being easy to plot because all the shots are on the contour lines.

**TRACE CONTOUR**

**Strip topos** are elongated topos used for street or sewer design. The center line of the strip topo is called the "P" line. This is the reason that the hand level is sometimes called a P-gun. The center line, or control line, is run first and points set at even stations

**STRIP TOPOS**

**Figure 8-6** Profile and cross sections

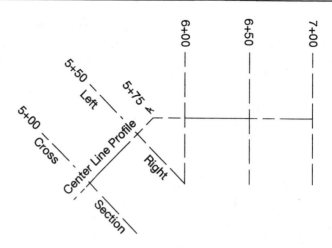

along the line and then at angle points. Levels are then run to establish the elevation of points on the control line. The side shots are then taken by one of several methods.

One method is to set up the instrument at each point and turn a 90° angle. The shots along the perpendicular line, or cross section, are then made in the usual way.

Another method is to use a right angle prism to give line and a cloth tape to measure the distance out from the P-line. Take the elevations with a level and rod.

Still another method is similar to the latter method, except that you take the elevations with a rod and P-gun. Elevations are given as a distance out from and plus or minus so many feet and tenths from the point on the P-line (10-ft. right, 4.7 ft. lower). Be careful of duplicating shots at the angle points or lag curves on the acute side of the angle while leaving a gap on the obtuse side (see Figure 8-6).

## PHOTOGRAMMETRY

When the area to be mapped is very large, or the terrain is difficult to cross, **photogrammetry** is used to make topographical maps. The American Society of Photogrammetry and Remote Sensing (ASPRS) defines photogrammetry as, "The art, science, and technology of surveying and measuring by photographic and other energy emitting processes." The role of survey technician in aerial photogrammetry is to set the ground control to be used by the photogrammetrist to create the finished map. You have probably seen the large white crosses painted on the paving along the highways. These are aerial **premarks** that are set to control the photo mosaic that is used to create a three-dimensional model. This model is used by the photogrammetrist to plot the contours of the finished map. The premarks also help the pilot to fly the correct flight line to take the photographs that are used to make

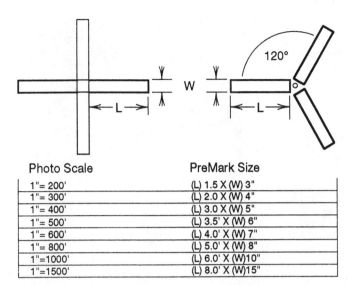

**Figure 8-7** Premarking for aerial photos

| Photo Scale | PreMark Size |
|---|---|
| 1"= 200' | (L) 1.5 X (W) 3" |
| 1"= 300' | (L) 2.0 X (W) 4" |
| 1"= 400' | (L) 3.0 X (W) 5" |
| 1"= 500' | (L) 3.5' X (W) 6" |
| 1"= 600' | (L) 4.0' X (W) 7" |
| 1"= 800' | (L) 5.0' X (W) 8" |
| 1"=1000' | (L) 6.0' X (W)10" |
| 1"=1500' | (L) 8.0' X (W)15" |

the mosaic. The location and size of the premarks are chosen by the company that will do the flying and photography. The scale of the finished map is used to set the flight height above the terrain and the flight height is used to determine spacing of the flight premarks. The size and the type of premark is determined by the flight height (see table in Figure 8-7).

The time between the setting of the premarks and the flight is critical. If the premarks are disturbed or wiped out for whatever reason, the flight must be repeated at some cost to everyone involved.

The premarks should be set with a great degree of care. The measurements of the premark targets are used by the photogrammetrist to find the exact center of the target. If these measurements are not correct, you may have to return to the point and reset the target or give the corrected measurements to your employer. This is embarrassing, at best.

When setting the premark, be sure to check that it can be seen from above without obstruction. In running the control, both horizontal and vertical, through the premark points, be sure to include the points in the traverse. Turn through them in the level run so that they will be included in the adjustments. An error of as little as a few hundredths will show up on analytical bridging of the photo mosaic.

Not all photo control points are premarked. Some are postmarked later or are picture points that are identified and pinpricked on the photograph in the field. A point for photo control should be on flat ground. A bush or tree standing alone, the intersection of two roads, the end of a sand bar in a stream, and the point where a sidewalk and back of curb meet are easily identified on a photo

5-19-78                                    NW-78  47C-10

**Figure 8-8** Aerial photo. *Courtesy of Washington State Department of Natural Resources*

and make good picture points (see Figure 8-8). The points should be pricked through the photo with a fine pin and identified on the back. Be sure to include coordinates and elevation or a number that will positively identify the point for the photogrammetrist. Topographic control requires that at least two horizontal and four vertical control points show on each photo in the overlap area between two adjacent photos. It is better to have too many points than too few, so if in doubt, set extra points. A field stereoscope is essential in locating picture points and identifying them on the air photo. After the topographical map is completed, it will need to be field checked by the survey party. This is done by running **profile** lines and comparing them to the contours on the map. Most topo maps are required to be accurate within one half of the contour interval of the finished map over 90% of the mapped area.

**GROUND CONTROL**

Once field data is collected and turned into the office, the notes are reduced and checked by a computer. If a data collector was in use on the total station, the information may be downloaded directly into a computer. There are several excellent software programs that reduce the raw field data to X-Y-Z coordinates that can be plotted or used to create a **digital terrain model**.

**DIGITAL MODELING**

Digital modeling is a method in which a three-dimensional (3-D) continuous surface is defined (modeled) by numerical data. The operator makes decisions about break lines, such as the top and toe of slopes, flow lines of creeks, and other physical features that affect the surface flow. Once the breaks are defined, the program allows for these and will not cross them with the triangles used for computing the contours.

After creating the triangle net, a grid is made from it. This allows 3-D views or calculating volumes by the **borrow pit** method. The points of equal elevation (contours) from the triangle net are computed by the program and plotted. These points of equal elevation are joined by a series of lines that show the contours of the surface. A contour-smoothing option will take out the angle points of the contour and make the smooth flowing contours associated with topo maps.

The operator must also identify all the culture such as property boundaries, roads, vegetation boundaries, trees, buildings, etc., and give a description for each. These features can then be shown as figures and be plotted and labeled on the finished map.

**COMPUTER-GENERATED TOPICAL MAPS**

The advantages of computer-generated topographical maps over hand-drawn maps are twofold. First is the ease of making changes without having to redraw the entire map. Second is the ability to create layers of information from the same data. One layer might show only the original ground elevations. Another could show the vegetation and still another, the culture. They must each be entered as sets if layering is planned.

Some computer programs even allow the plotting of a two-dimensional (2-D) drawing onto a 3-D projection of the surface

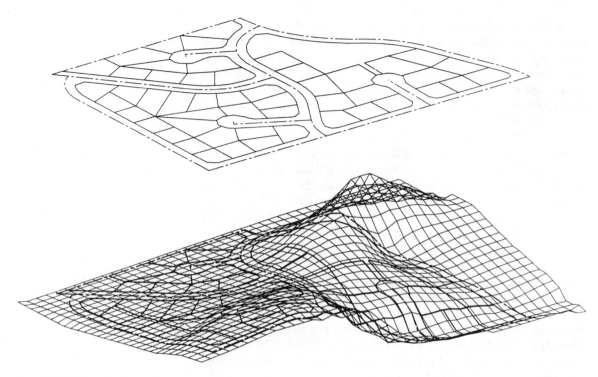

**Figure 8-9** 2-D overlay for a 3-D projection. *Courtesy of PacSoft*

(see Figure 8-9). This allows plotting of subdivisions, buildings, reservoirs, or other proposed improvements on a 3-D model on the video monitor. This makes changes possible without printing a hard copy of the map. Water flow and slope shading plots are also possible with the better programs. However, with the advantages come the disadvantages. GIGO is a computer term that stands for "garbage in, garbage out." If the operator enters faulty information, the computer cannot tell it from good data and will go right ahead as if it knew what it was doing. Do not become so impressed with the computer that you feel it cannot do anything wrong. There is still a need for a competent operator, no matter how sophisticated the computers and software become. Be sure to check all output for completeness and accuracy.

## A COMPETENT TOPOGRAPHER

On surveys that must be calculated and plotted by hand, the skill of the drafter is what separates a good map from a poor one. Just as the rodperson determines how good the field data will be, the person that interprets the field data and creates the finished map must be somewhat of an artist. Much practice and skill are needed to become a competent topographer; but once you have reached that level, there is a great demand for your services. There is a saying, "One picture is worth a thousand words." Try to visit the site of the project to see what the terrain is like. It is very helpful when making decisions about conflicting points on the map. If a visit to the site is not possible perhaps the field party can take some pictures to help you. If there is a doubt about the information shown in the field notes, the party chief must be questioned by the drafter. This may cause a delay in the project. A good topographical map is truly a team project and good communication skills are important for everyone involved.

Once the field notes are reduced, the elevations must be plotted on a sheet of paper to a suitable scale. A large protractor is essential for plotting stadia topo. The drafter might possibly construct a compass rose and place it under the drawing for use in laying out angles. The compass rose should be large enough in radius to allow the scale only to reach the circumference. This will allow the perimeter to be divided into units of at least 15 min. and interpolated to about 5 min., which is accurate for most topos.

You can also use an 8- to 12-in. protractor, but remember, the bigger the better. To use it, set the zero mark on the protractor on the backsight. Make a pencil mark at the degree mark that is nearest the horizontal angle in the field notes. Leave the pencil tip on the mark and set the zero end of the scale at the instrument station. Swing the scale against the pencil point. Move the pencil point to the distance on the scale and make a point on the paper. Use this point for the decimal point of the elevation and record the elevation of that point (see Figure 8-10). Be sure to lay off the

**PLOTTING A TOPO**

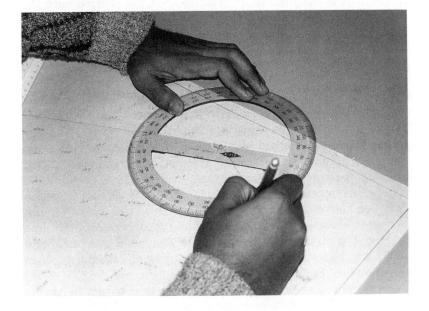

**Figure 8-10** Plotting a topo with a protractor

**Figure 8-11** Plotting a topo with a scale

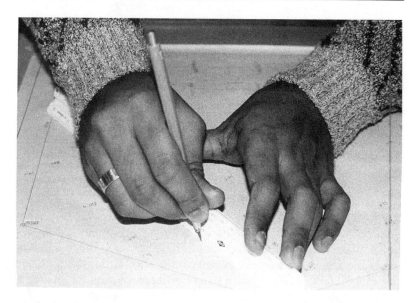

angle in the proper direction. Not all angles are to the right although it is easier to plot if they are. Keep this in mind when working in the field.

Put a check by each point in the field book as it is plotted to avoid missing any points. If the point has a description in the field notes, be sure to note it on the point (i.e., top, toe, f.l., etc.). After all the points are plotted, the points of equal elevation must be laid off between the plotted points. To do this, it is necessary to connect adjacent points with straight lines. As in the computer program, it is essential that break lines are not crossed with the triangles. Also, do not put diagonals between points of higher or lower elevation.

Once the triangles are determined, the points of equal elevation must be plotted. To do this, the proportional points along the diagonal must be calculated. This can be done on a calculator or graphically. With the calculator, it is a matter of proportion. The difference in elevation between the first point and the desired contour is divided by the difference in elevation between the two points. The result is multiplied by the scaled distance (any scale can be used) between the points. The distance to the contour line from the first point is given in scale units as the answer. If the distance from point 1 to point 2 is 0.75 inches, the difference in elevation between the two points is 8.2 feet. That makes the difference in elevation between point 1 and the desired contour line 3.6 feet. The distance to the contour is .33 inches. The remaining points are proportional to the scaled distance and the difference in elevation (see Figure 8-11).

Graphically, a scale and a 90° triangle are used to plot the points of elevation. The difference in elevation between the two points is

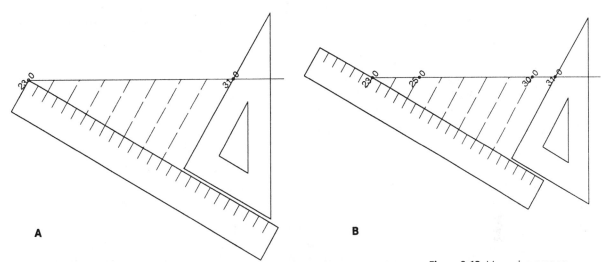

A

B

**Figure 8-12** Measuring proportional parts

calculated first. Then the zero point of the scale is set over the first point and a number of units equal to the difference in elevation are chosen on the scale. The total must be smaller than the distance between the two points in the same scale (see Figure 8-12A). Place the zero end of the scale on the lower elevation, then slide the scale along the triangle until the lower elevation corresponds to the units on the scale. Now you can lay off the even contours along the scale (see Figure 8-12B).

If the contours get too close together, as in a very steep slope, omit the intermediate lines and show only the main contours. On overhangs and caves, the contour may disappear inside the overhang but it is still there. These are shown as dashed contour lines.

On vertical surfaces, such as buildings, the contours are still there. They are stacked one on top of the other and are not visible; but they are still there. Contours crossing a stream bed or flow line point upstream or toward the higher elevation. Contours around a ridge point toward the lower elevation (see Figure 8-13).

The contours are identified by the elevation of the contour at the edge of the map, where they enter and leave from outside the mapped area. On closed contours, the elevation is placed in a break in the contour line. These breaks should be made in a place that will not cover a change in direction of the contour line (see Figure 8-14). On large, flat areas that do not have enough slope to be crossed by the contour interval, show the elevations as spot elevations with a plus (+) symbol to mark the spot. Print the elevation next to it.

Trees are shown and identified for size and type. Use a different symbol (see Figure 8-15) to show deciduous trees (trees that drop their leaves in winter) and coniferous (trees that are green year round).

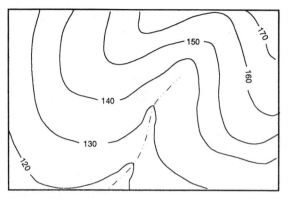

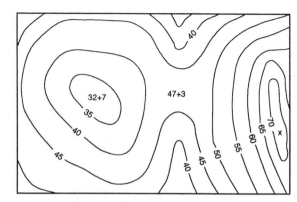

**Figure 8-13 (Left)** Contours crossing stream and ridges

**Figure 8-14 (Right)** Contour breaks for depression and peak

If you are going to crosshatch the buildings or stipple the concrete areas, turn the drawing over and work on the back of the sheet. By working on the reverse side, you need not be concerned about crossing the outline of the areas you are shading. If you use an erasing shield, you can clean up the edges or erase an area for lettering without harming the quality of your line work. When the blueprints are made, you cannot tell the difference. Also, smudges on the face of the drawing will be minimized.

Strip topos are usually made for elongated projects such as highway design or pipelines. The profiles and cross sections of the desired route are used the same as other topo data: to draw a plan or top view of the ground surface. However, for designing a road or pipeline, we also need to know what the surface looks like from a side view or profile. Profiles can also be taken from a regular topographical map by scaling a distance along the line of the desired profile between the contours. Plot the profile using the contour elevations crossed by the line.

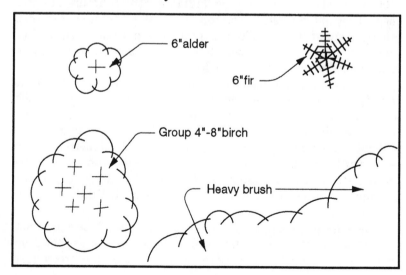

**Figure 8-15** Different types of trees

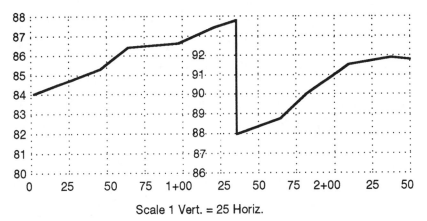

**Figure 8-16** Profile with vertical break

## THE PLAN AND PROFILE

A special type of paper is used to plot the plan and profile. The lower half of the paper has a grid for profiles printed on the reverse side of the paper. Be sure to have the grid on the back, lying face down on the drawing board so as not to erase lines if you must make changes. Notice the grid is in orange ink on the paper. This will print as dark lines on the blueprint if it is not erased. The upper portion is left clear and used as the plan portion. The grid area has heavier lines spaced every 10 ft. on a horizontal scale of $1'' = 40'$. It is divided into groups of five lines vertically to allow a vertical scale of $1'' = 4'$. The vertical exaggeration is needed to show the relief of the surface of the ground except in very steep terrain. The stationing of the plan and profile should be the same, except around curves. This allows the points to be brought straight down from the plan to the profile for ease in plotting. It is also an aid to reading the plans. If the elevations on the profile get too high or too low to plot, show a vertical break line. You can continue the elevation at a higher or lower point on the break line (see Figure 8-16).

## CROSS SECTIONS

Profiles at right angles to the main profile are called cross sections (see Figure 8-17). These are used to find the cross slope of the ground across the project.

On highway plans, the area between the proposed surface of the roadway and the natural ground is used to calculate material quantities for construction. The methods used to calculate these quantities are covered in Chapter 11.

## ALTA MAPS

Title companies require the most precise type of topographical map. They use these maps for research of the title to show any encroachments by adjoining property onto the property for which

**Figure 8-17** Cross sections for road

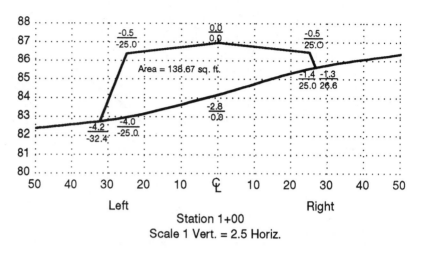

Area = 138.67 sq. ft.

Left

Right

Station 1+00

Scale 1 Vert. = 2.5 Horiz.

title insurance is sought. The maps also show any encroachment onto adjacent property by the property being mapped. The maps are known as ALTA maps. A copy of the requirements is available at the local office of the American Land Title Association (ALTA). Be sure to obtain one before beginning work in the field.

The boundary survey must show the measured bearings and distances as well as the record bearings and distances. All signs of lines of occupation surrounding the property should be shown along with their approximate age. This includes fences, walls, hedges, tree lines, and driveways. Any buildings close to the record boundary should be shown on the map; include the number of stories and type, such as one-story frame house. Be sure to check for basements or cellars that may extend over the property line. If walls of buildings are close to the line, check to see if the walls are plumb and record any encroachment to the nearest hundredth of a foot. If the walls are on the line, such as a party wall, record the thickness and type of construction material and whether it is a bearing wall or not. A bearing wall is one that supports the structure above it. Any building extensions such as eaves, ledges, cornices, chimneys, pipes, canopies, or windows that open outward should be shown with the amount of encroachment. All manholes, drain openings, signs of pipelines or buried cables (such as "do not dig" signs), or anything that might cross under the property must be shown. Show all power or utility poles around the property along with the pole number, type, and number of wires. An example is "3 power, 1 telephone, and 1 cable TV." One more time: BE CAREFUL AROUND POWER LINES! Some poles have guy poles or wires that extend onto the property. Be sure to locate these along with any wires that may pass overhead across the property. Any suggestion of water flow onto or from the property should be shown. Watch for drain pipes along the property line that might flow onto the property when it rains. The best way to measure the offsets to structures close to the line is to set the instrument up on the property line as surveyed. Sight along the line to another corner of the property. If the property line is obstructed, use an even foot offset line. Measure the distance to the object to be located at right angles to the property line with a tape or the level rod. Wave the tape slowly back and forth while the instrument operator records the lowest reading. This is like waving the rod for levels, only this time the rod is horizontal. The distance to the property corner is measured from the point on the property line at right angles to the object located. If the locations are measured with a total station, and angle and distance are shot to a prism, be sure to add the distance to the surface from the measurement point of the prism. All easements of record should be shown on the map along with their deed reference.

**PRACTICE IN PLOTTING CONTOURS**

To get practice in plotting contours, a grid topo is the easiest to do. Suppose that on the tract of land that was traversed in the last chapter (see Figure 7-9) a grid topo was made using corner number 1 as the starting point. A 100-ft. grid was run to the south and west perpendicular to the property line between corners 1 and and 8. Points were set out to 1100 ft. to the west and 1000 ft. south along the property line. The property is free of trees and is gently sloping to the east and south. The instrument was set up on the grid points along the property line and a 90° angle was turned. Points were set using a prism pole and the stake out routine from the instrument. A lath was set at each point and given a code distance out and over. The first point at corner 1 was marked 0s0w. The next point on the north line was marked 0s1w, the next 0s2w, the next 0s3w, and so on to 0s11w at 1100 ft. from the corner. The next line to the south was marked 1s0w, 1s1w, 1s2w, out to 1s11w. Each of the succeeding lines was marked in the same way up to 10s0w. This line needed to be run out only 400 ft. to get all the points needed. It was run only to 10s4w. An assumed elevation was used, with an elevation of 100 ft. being assigned to corner number 8, it being the lowest corner. The following elevations were read at the grid points:

0s0w-125.2, 0s1w-126.5, 0s2w-128.2, 0s3w-129.6,
0s4w-130.8, 0s5w-132.3, 0s6w-134.1, 0s7w-136.0,
0s8w-140.3, 0s9w-144.7, 0s10w-147.2, 0s11w-150.6,
1s0w-122.3, 1s1w-124.2, 1s2w-126.0, 1s3w-127.5,
1s4w-129.6, 1s5w-131.8, 1s6w-133.9, 1s7w-136.3,
1s8w-140.6, 1s9w-144.2, 1s10w-147.0, 1s11w-151.2,
2s0w-119.0, 2s1w-121.0, 2s2w-123.1, 2s3w-124.9,

2s4w-127.0, 2s5w-130.8, 2s6w-133.5, 2s7w-136.7,
2s8w-140.9, 2s9w-144.0, 2s10w-146.5, 2s11w-156.7,
3s0w-116.0, 3s1w-118.0, 3s2w-120.2, 3s3w-122.2,
3w4w-124.0, 3s5w-128.9, 3s6w-132.7, 3s7w-137.3,
3s8w-141.0, 3s9w-143.3, 3s10w-146.3, 3s11w-150.5,
4s0w-113.1, 4s1w-115.0, 4s2w-117.2, 4s3w-119.0,
4s4w-121.8, 4s5w-126.0, 4s6w-130.7, 4s7w-135.0,
4s8w-138.2, 4s9w-140.8, 4w10w-141.5, 4s11w-143.7,
5s0w-110.0, 5s1w-112.3, 5s2w-114.3, 5s3w-116.0,
5s4w-118.5, 5s5w-122.2, 5s6w-126.7, 5s7w-130.0,
5s8w-133.0, 5s9w-136.0, 5s10w-137.5, 5s11w-138.4,
6s0w-107.1, 6s1w-109.0, 6s2w-111.0, 6s3w-113.0,
6s4w-114.7, 6s5w-118.0, 6s6w-122.0, 6s7w-124.8,
6s8w-127.0, 6s9w-129.5, 6s10w-131.5, 6s11w-133.0,
7s0w-104.0, 7s1w-106.0, 7s2w-107.0, 7s3w-110.0,
7s4w-112.3, 7s5w-115.0, 7s6w-117.8, 7s7w-120.0,
7s8w-122.2, 7s9w-124.0, 7s10w-125.5, 7s11w-127.3,
8s0w-101.2, 8s1w-103.0, 8s2w-105.8, 8s3w-108.4,
8s4w-110.5, 8s5w-112.3, 8s6w-114.8, 8s7w-117.0,
8s8w-119.0, 8s9w-120.5, 8s10w-122.0, 8s11w-123.9,
9s0w-98.1, 9s1w-101.3, 9s2w-104.1, 9s3w-106.5,
9s4w-107.8, 9s5w-110.7, 9s6w-112.5, 9s7w-114.6,
9s8w-116.2, 9s9w-118.0, 9s10w-119.5, 9s11w-121.0,
10s0w-95.0, 10s1w-99.0, 10s2w-102.0, 10s3w-104.5,
10s4w-106.7.

On a 18 × 24 sheet of paper, plot out the points and label the elevations. Use the largest scale possible. You will be using the topo for designing and plotting a subdivision in Chapter 14. Establish the contour lines between the topo points and draw them in. Use a heavy line for the 10-ft. contours and a medium line for the 5-ft. contours. Then draw in the 1-ft. contours, using a broken line. Show the elevations of the 5- and 10-ft. contours at the ends of the contour lines.

# Review Questions

1. Why should the prism be set to the HI of the instrument on topo?
2. What is the signal to take a rod shot or to read the rod?
3. What is the signal to raise the rod or prism for a higher shot or "raise for red"?
4. Where should topo shots be taken to show changes in elevation?
5. What is "culture" on a topo?

6. What information should be given about trees?

7. What shots should be taken on a street cross section?

8. What shots should be taken on a manhole or catch basin?

9. Why should you not use metal-faced rods around power lines?

10. Why should you show paths and evidence of travel across the property?

11. When would you use a "grid topo"?

12. When would you use the "trace contour" method of topo?

13. What type of topo would you use for long narrow topos, such as streets or pipelines?

14. If the area is large or the terrain is difficult, what would be a good way to get a topo of the property?

15. What size of premarks should be set for photo control?

16. Why should you be very careful when setting out premarks?

# Survey Control

This chapter does not dwell on the geodetic control points set by the National Oceanic and Atmospheric Administration (NOAA) or its offices, the National Ocean Survey (NOS) and the National Geodetic Survey (NGS). These surveys are designed to control the position and elevation over the vast distances of the United States.

The newest adjusted datum is the North American Datum of 1983 (NAD 83), the North American Vertical Datum of 1988 (NAVD 88), and the NAVSTAR Global Positioning System (GPS). Information on these can be obtained by writing to:

National Geodetic Information Branch
N/CG174, Rockwall Building, Room 24
National Geodetic Survey, NOAA
Rockville, MD 20852
Telephone: 1-301-443-8631

The NAD 83 data record contains:

The NAD 27 station identifier
The station name
The latitude and longitude
The northing and easting in meters
The plane coordinate zone
The convergence mapping angle
The scale factor at the station
The elevation in meters
The **geoid** height in meters
The positional quality of the point on a 1 to 4 scale

## MONUMENTS OF RECORD

These points are the main control points that you will be using along with the standard corners of the public land system (section and ¼-section corners) and the monuments of record set by local registered land surveyors and city, county, and state surveyors. Information on local survey monuments can be obtained from several sources. The tax assessor office has tax records on all property in the county along with a description that will allow you to find the Record of Survey map at the local registrars office or from the city or county engineers office. Information on the field notes, plats, maps, and other papers relating to the Surveys of the Public Lands of the United States are available for each state. You will find them listed in Appendix B at the back of this book.

Information on center line monuments found in the streets adjoining the project can be obtained from the road or transportation department of the city, county, or state. If you find property markers with RLS or PE numbers stamped on a brass tag or plastic cap, information on the marker can be obtained from the surveyor or engineer that set the point. Once the boundaries of the project are established and any missing points are reset and checked, the project control can be established.

A decision must be made on how accurate the work must be before starting on the control traverse in the field. The standards of accuracy for various projects have been established by the Federal Geodetic Control Committee (FGCC) and are available from NOAA. The American Congress on Surveying and Mapping (ACSM) has established four general classes of **cadastral** surveys (a survey of the boundary of the public lands). These are

1. Class A-Urban Surveys: surveys of land lying within or adjoining a city or town. This would include the surveys of commercial and industrial properties, condominiums, townhouses, apartments, and other multiunit developments, regardless of geographic location.
2. Class B-Suburban Surveys: surveys of land lying outside urban areas. This land is used almost exclusively for single family residential use or residential subdivisions.
3. Class C-Rural Surveys: surveys of land such as farms and other undeveloped land outside the suburban areas that may have a potential for future development.
4. Class D-Mountain and Marshland Surveys: surveys of lands that normally lie in remote areas with difficult terrain and usually have limited potential for development.

Instruments and methods capable of the precision needed must be chosen to achieve exactness required. Precision is the ability to measure something repeatedly and have all the measurements fall within the area of probability for the true measurement. It relies on the combination of instrument quality and the ability of the operator. Instrument manufacturers test their instruments to determine the standard deviation from true for each type of instrument. This tells the operator how much precision can be expected from a properly adjusted instrument. The ability of the operator can be judged only from past performance. Only perfect practice makes perfect.

A certain precision can be expected with the proper equipment and operator. This, however, does not rule out **errors**. There are two types of errors that can be expected in any measurement; **systematic** (cumulative) and **accidental** (random). These are always present and can be corrected for in the final adjustment. A **blunder** cannot be expected, or corrected for, in any normal mathematical method. A blunder is a mistake, such as transposing two numbers (87.69 for 87.96) or reading the wrong foot on the rod with the tenths and hundredths correct. Only careful work can prevent blunders. If you feel there has been a blunder made, call it to the attention of the other crew members so that a check can be made and the correct information recorded. No one has ever been fired for catching a possible mistake and correcting it before it causes damage. Remember, it is not a mistake until the concrete is poured.

Systematic errors are caused by known deviation from normal, such as temperature on a taped distance, curvature and refraction in leveling, or an index error in the vertical reading. All systematic errors must be found and corrected.

Random errors tend to be small and are just as liable to be too big as too small. They are referred to as "plus and minus" and, as such, tend to cancel each other out. They do not accumulate as systematic errors do and are usually caused by rounding off a measurement, such as reading 100.053 as 100.05 or 79-45-24.7 as 79-45-25. Remember that errors in alignment cause the measured distances to be longer than straight line distances and are not compensating. When reading the hundredths on a tape, if in doubt, try to round down. On a rod reading, if the rod is not plumb, the reading will be high. The rule still applies; try to round down when possible.

When measuring a single direction (distance or angle), the laws of probability state that 68.3% of all measurements will fall within the standard deviation(s). This law also applies to the root mean square of the deviations from the mean. Almost 99.7% of all measurements will be less than 3 times $s$ ($3s$). A $3s$ error can be expected to occur only once in 370 measurements. The $3s$ error for a set of six direct-reverse readings with a 6-second theodolite is 4.7 seconds. The amount that error propagates (accumulates) is found by the formula:

$$E_p = \sqrt{(E_1{}^2 + E_2{}^2 \ldots + E_n{}^2)}.$$

In a 6-sec. instrument, the "maximum" error that would be expected in a 10-station traverse would be $E_p = \sqrt{10\,(4.7)^2} = 14.9$ seconds. If an error greater than 14.9 seconds accumulated in the traverse, a blunder would be suspected. When you figure in the error for distance for an EDM with a standard deviation of ±5mm + 5ppm or

$$E_d = 3\sqrt{(5mm/(25.4 \times 12)^2 + (5d/10^6)^2}\ (d = \text{distance in feet}),$$

you begin to get an idea of your probable closure error. Suppose this error is not within the maximum allowed for the class of survey. You must either get more precise instruments or find a way to use fewer stations. One possibility is using GPS equipment for greater accuracy.

An EDM should be checked for accuracy on a known baseline before starting on a control traverse. A list of calibration base lines for each state is available from NOAA. Be sure to peg test the level you will be using for vertical control at least at the beginning and end of each week, if not each day.

## THE POSITION OF THE MONUMENTS

The position of the monuments to be set on the control line should be carefully chosen. They must be set in a secure position away from construction disturbance but close enough to be of use. The type of monument depends on the conditions found on the job site. If the ground is soft, an iron rebar can be driven into the ground below subgrade and referenced outside the construction area. A 2 × 2 hub driven below the surface of the ground and surrounded by a lath basket with florescent flagging laced through the lath is a standard tie. Be sure to explain to the construction superintendent the importance of the ties and the monuments. Extra ties set in holes and not marked, except in the field book, are good insurance. There is an old story that a surveyor cannot become lost; all he has to do is set a hub and wait for the bulldozer to come by and knock it out—then he can catch a ride out with the driver. It is also good practice to drive the hubs in vertically so that, if they are hit, you will notice.

## TYPES OF TIES OR REFERENCE POINTS

There are two types of ties or **reference points** generally used: **swing ties** (see Figure 9-1) and **throw-over ties** (see Figure 9-2). Swing ties can be set with the tape alone and are handy to relocate using only a tape. Throw-over ties require the use of an instrument and tape.

A good practice is to set two ties at about a 90° angle apart and on an even foot, such as 25.00 or 40.00 foot. This allows the relocation of the monument by using only a tape and, by knowing the approximate location (within a foot), being able to find the point without having exact knowledge of the tie distance. If the even-

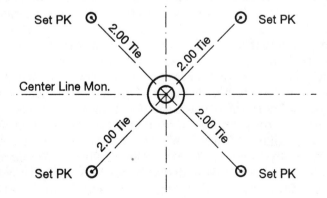

**Figure 9-1** Swing ties to center line monument

**Figure 9-2** Throw-over (plunged) ties to center line monument

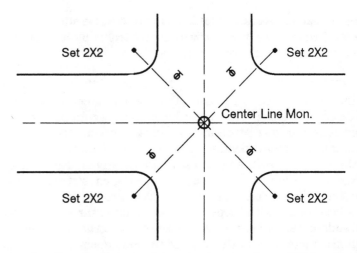

foot swing ties are then used to set up on, and an additional stake is set at the same distance on a throw-over, the swing ties become throw-over ties. This way, if one of the tie stakes is knocked out or disturbed, the others can still be used.

The throw-over stakes can sometimes be put in holes and not flagged so that the only record of them is in the field book. This insures against tampering by third parties. It happens, on occasion, that some parties object to the project that you are surveying and maliciously destroy or, even worse, move the survey stakes, hoping to stop progress of the job. The best way to find this tampering is by setting extra reference points (RPs) known only to the surveyors. Do the same thing for bench marks on the job. Set extra auxiliary BMs and check them if your levels suddenly do not close. The BMs might have been disturbed.

## CONTROL POINTS

Always use extra care when setting control points. A good surveyor knows when to be precise and when to just be fast, or what's called "down and dirty surveying." All angles should be called out to the recorder and repeated back to the instrument operator to be sure that numbers are not transposed in recording. The newer total stations that use data recorders help prevent transposition of data. Distances on the EDM should be repeated and the average reading used. For **control surveys**, mount the prism on a tripod to assure accurate reading. Be sure the prism is plumbed over the point, pointed at the instrument and the HI measured, if necessary.

Use a target to check the rod reading to the nearest thousandth of a foot or millimeter on control level circuits. Be careful of using bolts on fire hydrants for TBMs. Occasionally, a riser section of pipe below the hydrant is added or removed, which changes the elevation of the hydrant. Drive spikes horizontally into the trunk of a tree and not vertically into the top of the roots. A tree grows upward from the top and a spike in the trunk will not change elevation, but a spike in an exposed root will be forced out as the root grows in circumference. Drive the spike just above the root swell so that, if the tree should grow over the spike, the likelihood of a chainsaw hitting it is lessened if the tree is cut down. Driving spikes into a telephone or power pole is not a safe practice. The person climbing the pole can get a leg caught on the protruding spike and be injured or fall.

On some very large projects, the control points will be used as control for aerial mapping. In this case, the location of the points are planned in coordination with the photogrammetrist. Tall buildings or trees close to the point could obscure the premark from the camera. Additionally, tall objects often cast shadows that may obscure the premark. A **cone of visibility** of 45° in all directions is advised by most aerial photography companies. The view of the sky is also important if GPS is going to be used. For GPS, choose a site with a clear view of the sky above 15° from the horizon. Also try to choose a point that will allow location of the GPS receiver within 100 ft. or 30 m. of the control point. GPS allows the use of control points that are not intervisable but, remember that a project control must have at least two intervisable points for azimuth unless an astronomical azimuth is used. Choose a point with good all around visibility if radial stakeout of the project points is planned.

After the field control is completed and checked for blunders, the notes or data should be given to the office technician who does the calculation. Good field notes require no further explanation but it is a good idea for the party chief and the office tech to go through the notes together just to be sure there is no misunderstanding. Most survey offices use computers to reduce the raw field data to X,Y,Z coordinates for use with a CO-GO (coordinategeometry) program to design the finished project. With the increased need for accurate positional tolerance on most surveys, the least squares adjustment comes into use more and more. Once the traverse is closed and adjusted, any undisclosed blunders should have been found. The office tech and the party chief go over any discrepancy and make any necessary field changes. The office tech then receives the revised information and enters it into the computer. The field information is used to compute the unadjusted coordinates of each point on the control survey.

A knowledge of the adjustment of survey data is most useful for the office tech who wants to move ahead in the survey business. You should obtain a copy of *Analysis and Adjustment of Survey Measurements* by Edward M. Mikhail and Gordon Gracie, published by VonNostrand Reinhold, 115 Fifth Ave., New York, NY

**Figure 9-3** Field computer. *Courtesy of Corvallis Microtechnology, Inc.*

10003. The use of matrix algebra is introduced and used for adjusting survey data. Probability is explained and used to show how to develop the most probable position of survey points. Least squares adjustment is thoroughly explained and used in adjusting the relative position of control.

A least square adjustment program, Star*Net, is available from Starplus Software, Inc., 460 Boulevard Way, Oakland, CA 94610. The "Star*Net Least Squares Adjustment Program" is easy to use, even for the inexperienced office tech, and will greatly reduce the fear of "most probable position" calculations. Control traverses and level loops can be adjusted mathematically and the ellipses error is shown for each point. Once the adjusted coordinates of each control point have been computed and given to the field tech, the secondary control can be set using radial stake-out programs.

There are several excellent calculators available for the field (see Figure 9-3). These calculators or, more accurately, hand-held computers use survey programs to simplify the field computations. The need for more accurate surveys becomes greater as the value of property increases. With the use of the tools now available at modest cost to the surveyor, that surveyor can set points with greater accuracy and precision than was possible in the past.

# Review Questions

1. What do the following initials represent?
   a. NOAA
   b. NOS
   c. NGS
   d. NAD
   e. GPS
   f. NAVD
2. Define precision.
3. What are the two types of errors expected in measurements?
4. Define each of the errors in question 3 and give an example of each.
5. What is a blunder?
6. What should you do if you detect a blunder?
7. What is the expected error for a 6-sec. theodolite in 10 setups?
8. How often should you peg test the level?
9. Give an example of swing ties.
10. Give an example of throw-over ties.
11. What is the "cone of visibility" for an aerial premark?

# Horizontal and Vertical Curves

**10**

## Objectives

After completing this chapter and the field practice, the student should be able to:

1. Define the four types of horizontal curves.
2. Identify the elements of a simple horizontal curve.
3. Compute the layout of a simple curve, from the PC, at even stations.
4. Lay out a simple curve using a transit/theodolite and tape or EDM.
5. Define two types of vertical curves.
6. Given $G_1$, $G_2$, and the length, compute a vertical curve.

---

The basic layout of a modern highway is a series of straight sections joined together by curved sections. If the curved sections were not put in to join the straight sections, the change of direction would be too abrupt and the road would be unsafe for travel with any speed. By designing the curves to accommodate the desired maximum speed of the vehicles using the road, a comfortable transition from one straight section to the next occurs. The straight sections are known as "tangents." The curves are either **simple horizontal curves** or a combination of transitional **spiral** curves and circular curves. A spiral curve helps to gradually change the vehicle's direction from the tangent to the circular curve and back to the next tangent. Spirals are also used to change the flow of

**THE CHANGE OF DIRECTION**

water in flood control channels to prevent excessive erosion of the banks. Railroad curves are always laid out with spirals to prevent derailment of the wheels caused by too rapid a change of direction.

**TYPES OF HORIZONTAL CURVES**

There are four types of horizontal curves:

1. **Simple** Curves. The simple curve is an arc of a circle. The radius of the circle determines the sharpness or flatness of the curve (see Figure 10-1).
2. **Compound** Curves. Sometimes the terrain makes it necessary to use a compound curve. A compound curve is two simple curves of different radius joined together and curving in the same direction (see Figure 10-2).
3. **Reverse** Curves. A reverse curve is two simple curves joined together but curving in opposite directions. This curve is not safe because of the violent change of direction and should have a short piece of tangent inserted between the curves if possible (see Figure 10-3).

**Figure 10-1** Simple curve

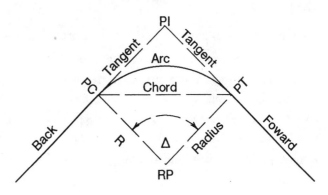

**Figure 10-2** Compound curve

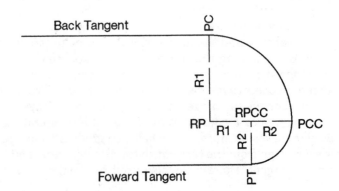

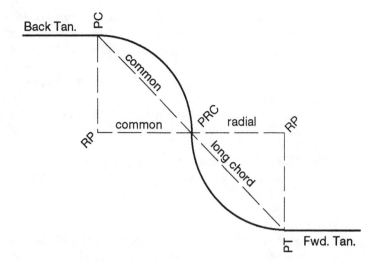

**Figure 10-3** Reverse curve

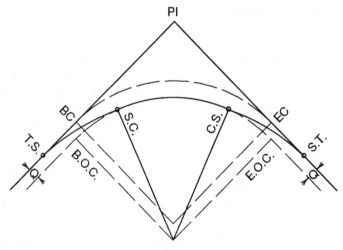

**Figure 10-4** Spiral curve

4. **Spiral Curves.** Sometimes called the transitional spiral curve, a spiral curve has a constantly changing radius. The spiral is used to provide a gentle transition from the tangent to a simple curve or to join two simple curves in a compound curve (see Figure 10-4).

---

The elements of a simple curve are

**PI** The point of intersection of the back and forward tangents.

**I** The intersecting angle. The deflection angle between the back and forward tangents when occupying the PI. The intersecting angle and the central angle are the same angle.

**ELEMENTS OF
A SIMPLE CURVE**

**R** The radius of the circle of which the curve is an arc PC (point of curvature) or BC (beginning of the curve). The point where the back tangent meets the curve is 90° to the radius point.

**POC** A point on curve. Any point along the curve between the PC and the PT.

**PC or BC** The point of curvature or beginning of curve. The point where the back tangent meets the curve and the radius is 90° to the tangent.

**PT or EC** The point of tangency or end of curve. The point where the curve meets the forward tangent and the radius is 90° to the tangent.

**L** The length of curve measured along the arc from the PC to the PT.

**T** The tangent distance or the distance from the PC to the PI and from the PI to PT, measured along the back and forward tangents of the curve.

**Δ** Delta or the central angle at the radius point. The same value as angle I (delta).

**LC** The **long chord** from the PC to the PT or a straight line connecting the beginning and end of a simple curve.

**E** The external distance. The straight line distance from the PI to midpoint on the curve (mid-POC).

**M** The middle ordinate or midordinate. The straight-line distance from the midpoint on the curve to the middle of the long chord (LC). A line from the PI to the radius point passes through the mid-POC and the midpoint of the LC and bisects the delta angle. It is 90° to the bisector of I.

**D** The degree of curve defines the sharpness or flatness of the curve. There are two types of degree of curve.

1. Degree of curve, "arc definition," states that the degree of curve is the angle formed by two radii drawn from the radius point of a circle to the ends of an arc 100.00-ft. long. In this definition, the degree of curve and the radius are inversely proportional. For example,

$$D : 360° : : \text{Arc} : \text{Circumference}.$$

If we substitute $D = 1°$, we get $1 : 360 : : 100 \text{ ft.} : 2\pi R$ or $1 : 360 : : 100 : 6.2831853 R$. Solving for R we get $36000/6.2831853 = 5729.58$ ft. If the D were 5°, the radius would be 1145.92 ft. The arc definition is used on highways where the stationing follows the curve.

2. Degree of curve, "chord definition," states that the degree of of curve is the angle formed by two radii drawn from the

radius point of a circle to the ends of a chord 100.00-ft. long. The radius is computed by using the formula

$$50/\sin \tfrac{1}{2} D = R.$$

Using 1° as D and substituting, we get $50/\sin 30' = 5729.65$ ft. For a 5° curve chord definition, the radius is 1146.28 ft. The chord definition is used on railroads.

**HORIZONTAL CURVE LAYOUT**

On curves with long radii it is not practical to stake the curve by using the radius point and swinging the radius to lay out the curve with a steel tape. It is much easier to set the instrument on the PC and turn deflection angles from the back tangent and pull the **short chords** to the station points or delta layout points (see Figure 10-5).

On some curves, the contractor will want to even stations set on the curve. On others, the curve will be broken into even parts or deltas (¼, ½, ¾) on short radii curves. The distance between stakes will depend on the percentage of grade and the length of curve. Commonly used chord lengths are:

0° to 3° curve—100 ft.
3° to 8° curve—50 ft.
8° to 16° curve—25 ft.
over 16° curve—10 ft.

Deflection angles to the stakes are throw-over (deflected from the extension of the back tangent) angles from the tangent. The total of the deflection angles to the short chord stakes should equal half of the central angle. The last chord should equal the long chord of the curve and the deflection angle should be a half delta.

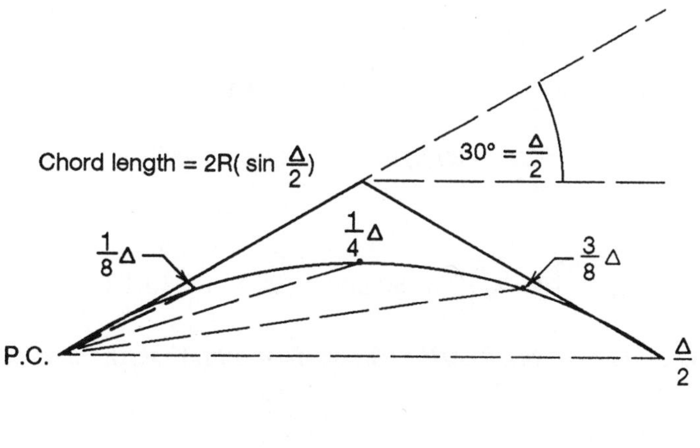

**Figure 10-5** Horizontal curve layout

**FORMULAS FOR
SIMPLE CURVES**

Some commonly used curve formulas for simple curves are:

$$R = \frac{5729.58}{D} \text{ (arc definition)}$$

$$R = \frac{50}{\sin \frac{1}{2} D} \text{ (chord definition)}$$

$$T = R \tan \frac{1}{2} I$$

$$L = \frac{I}{100\ D} \text{ (Exact for the arc definition)}$$

$$PC = PI - T$$

$$PT = PC + L$$

$$E = T \tan \frac{1}{4} I$$

$$M = R - (R \cos \frac{1}{2} I)$$

$$LC = 2R \sin \frac{1}{2} I$$

$$I \text{ (in radians)} = L/R$$

Three elements must be known to solve a simple curve. Two of the elements are usually the PI and the central angle. The third element will usually be the degree of curve or the radius. The PI and delta angle are normally determined on the preliminary survey and are designed to fit the terrain and design speed of the road. Assume the following are given:

$\Delta = 75°$, PI 18 + 00 and D = 15°.
Arc Definition

$$R = \frac{5729.58}{15} = 381.97 \text{ radius}$$

$$T = 381.97 \times 0.076733 = 293.10 \text{ tangent}$$

$$PC = 1800 - 293.10 = 15 + 06.90 \text{ PC}$$

$$PC + L = PT$$

$$L = 100 \times \frac{75}{15} = 500.00 = 20 + 06.90 \text{ PT}$$

$$LC = 2 \times 381.97 \times 0.60876 = 465.06 \text{ long chord}$$

If the curve is to be staked on stations and half stations, the distance to the first stake is

$$15 + 50 - 15 + 06.90 = 43.10.$$

The length of arc divided by the radius (L/R) gives the delta angle to the stake:

43.10/381.97 = 112,836 radians, converted to degrees = 6.4650°, converted to deg. min. sec. = 6°27′54″.

Store your decimal degrees so they can be used to calculate the half delta angle and the total delta for a check. For the remaining lengths of arc the calculations are:

[50/381.97 = .130900 radians = 7.5000 degrees = 7°30′00″]
[20 + 06.90 − 20 + 00 = 6.90]
[6.90/381.97 = .018064 radians = 1.03500 degrees = 1°02′06″]

There are nine (9) 50-ft. stations plus the starting and ending short stations in the curve. Add the total station deltas to see if they add up to the total delta angle.

$$6.4650 + (9 \times 7.5000) + 1.03500 = 75.0000 \text{ check.}$$

To calculate the short chords use twice the radius times the sine of half delta.

$$2 \times 381.97 \times \sin 3.2325 = 43.0769 \text{ first chord} = 43.08$$

To get the half delta, recall the stored delta and divide by two.

Recall, 6.4650/2 = 3.2325
$$2 \times 381.97 \times \sin 3.7500 = 49.9641.$$

The next nine chords are 49.96 ft.

$$2 \times 381.97 \times \sin .5175 = 6.8999.$$

The check in chord should be 6.90 ft.

Another way to stake out the curve is to measure all the distances from the PC with an EDM. The only difference in the calculations is that the chord lengths would all be from the PC and the deflection angles would be additive.

The first angle would be 3°13′57″, the next angle would be 3°13′57″

+ 3°45′00″ = 6°58′57″. The following angle would be 6°58′57″ + 3°45′00″ = 10°43′57″ and so on to the last 0°31′03″ for a check in total of 37°30′00″.

To calculate the chords use 2 R sin ½ delta = LC.

$$
2 \times 381.97 = \begin{array}{l}
763.94 \times \sin 3\text{-}13\text{-}57 = 43.08 \\
763.94 \times \sin 6\text{-}58\text{-}57 = 92.87 \\
763.94 \times \sin 10\text{-}43\text{-}57 = 142.26 \\
763.94 \times \sin 14\text{-}28\text{-}57 = 191.05 \\
763.94 \times \sin 18\text{-}13\text{-}57 = 239.02 \\
763.94 \times \sin 21\text{-}58\text{-}57 = 295.96 \\
763.94 \times \sin 25\text{-}43\text{-}57 = 331.68 \\
763.94 \times \sin 29\text{-}28\text{-}57 = 375.98 \\
763.94 \times \sin 33\text{-}13\text{-}57 = 418.67 \\
763.94 \times \sin 36\text{-}58\text{-}57 = 459.56 \\
763.94 \times \sin 37\text{-}30\text{-}00 = 465.06
\end{array}
$$

Check LC

## PROCEDURE TO LAY OUT A SIMPLE CURVE

The usual procedure to lay out a simple curve in the field is

Step 1. With the instrument set up on the PI, the instrument operator sights the back tangent, the previous PI or PT, with the instrument reversed. While sighted on the back tangent, the PC is put in at the calculated distance from the PI (T). The deflection angle is then turned (I or delta) and the PT is set at the same distance (T). Now, set the next PI on the forward tangent at the correct distance.

Step 2. While the tapeperson returns to the instrument, the instrument is set up on the PC. Leave a sight on the PI facing the PC and another on the PT facing the PC, if possible, for a clear line of sight. The angle between the PI and the PT is turned and checked to be certain it equals half delta.

Step 3. With 0°00′00″ set on the PI, the first deflection angle is laid off (3°13′57″) and the chord measured (43.08). If long chords are used, the tapeperson moves on to the next point. If short chords are used, the rear tapeperson moves ahead, after setting a lath marking the station. An arc is scribed on the ground at the approximate point for the next station.

Step 4. The next deflection angle is laid off (6°58′57″). Either the long chord (92.87) is used to set the point on the new line or the line is given along the scribed arc and a stake set at the point of intersection. The distance is set from the last station (short chord) or with the EDM from the PC (LC) and a lath set to mark the station.

Step 5. The succeeding stations are laid out in the same manner. The last station being the PT for a check. To make it faster to get the proper first point, when moving ahead for the head tape, and after the second full station is set, sight back along the curve and get the approximate offset from a straight line through the two stakes to the curve. Use this distance to line in on the rest of the stations. It should not be necessary to move more than twice to get a line and distance to set the hub.

Sometimes it is not possible to see all the points on a curve from the PC. It may be necessary to move ahead to one of the points on the curve or it may be easier to move ahead to the PT and back in the remaining points. To move ahead to one of the stations, move the instrument ahead to the last station set. Set the deflection angle for that station in the instrument so that the scope will be tangent to the curve at 0°00′. Continue to set the deflection angles already calculated for the curve as if still at the PC (see Figure 10-6).

Sometimes, it is faster to set up on the PT and back in the curve. The instrument is already set up on the forward tangent and the tangent stations can be set without having to move the instrument ahead. The fewer setups made, the more time saved.

When calculating curves in the office, it will be much appreciated by the field crew if the information for the curve includes the central angle, the radius, the arc length, and the tangent. These are given on most computer program printouts, so extra work is not necessary to get the information needed. Sometimes, the field crew

**Figure 10-6** Obstructed line of sight

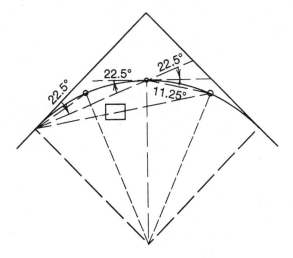

lacks the time to calculate them on the job site with equipment operators breathing down their necks. Cooperation between the field and office can make the whole job run smoother.

## COMPOUND CURVE

Compound curves are two simple curves joined together with the PT of the first curve becoming the PC of the next curve. In this case, instead of the point of compound curvature being called both PC and PT, call it PCC. It is easiest to lay out a compound curve on the PCC and back in the first curve. Then go directly into the second curve without moving the setup. The forward tangent of the first curve lays on the same line as the back tangent of the second curve. If the half delta of the first curve is set in the instrument and sighted on the PC of the first curve, 0°00′00″ will be along the common tangent. The second curve can be laid off without resetting.

## REVERSE CURVE

A reverse curve is two simple curves joined together with the PT of the first curve becoming the PC of the second. However, the second curve curves in the opposite direction, or the reverse, of the first curve. The common PT-PC becomes PRC (point of reverse curvature) for the two curves. A line drawn from the radius point of the first curve passes through the PRC and the radius of the second curve. Also, a line drawn from the PI of the first curve to the PI of the second curve passes through the PRC. This gives both curves a common tangent, as in a compound curve. The instrument can be set up on the PRC and the first curve backed in while the second is laid out normally. It is good practice, on a road or highway design, to put a short piece of tangent between the two curves to allow the driver of a vehicle a short time to center the wheel before entering the second curve. A reverse curve is a good way to join two parallel lines (see Figure 10-3).

As surveyors increase their use of computers and total station instruments, more use will be made of the radial stake-out feature of most total stations to lay out curves. With a radial stake-out program, all the points on any curve can be set from any control point. Be sure to make independent checks of random points to be sure that there is no error in alignment when doing radial stake outs.

## OFFSETS

In heavy timber it may be necessary to use **tangent** or **chord offsets** to set the points on the curve. A knowledge of coordinate-geometry (CO-GO) is very useful for these methods. Typical CO-GO programs are bearing/bearing intersections, bearing/distance intersections, and distance/distance intersections. Most useful for tangent and chord offsets is the "offset from a point to a line," given the bearing of the line and the coordinate of the point on the curve.

Pick a line through the timber that requires little brushing. Using that line as the bearing, run the offset program. This will give a distance along the cleared line and a distance at right angles to the line to set the point on the curve. A right angle prism will simplify the setting of the POC from the base line. This method cuts down considerably on the amount of brushing required and is a great time saver.

## TRANSITION SPIRALS

Transition spirals are used to gradually change the direction of travel of a body in motion. Whether that body is a truck, a car, or water, it tends to continue in the same direction. If the direction of travel is changed abruptly, centrifugal force tends to keep it moving in the same direction. If a curve of gradually increasing sharpness is inserted between the tangent and the curve, the body is gently eased around the change in direction. On a properly designed spiral curve, the steering wheel may be held in one position and the spiral curve along with superelevation of the roadway will take the vehicle through the turn with little or no change in steering wheel position. This creates a much safer highway for the driver and is also easier on the mechanics of the vehicle.

The radius of a spiral curve decreases from infinity, at the tangent to spiral, to be equal to the simple curve radius at the spiral to curve point. Conversely, the radius increases from a simple curve radius at the curve to spiral and back to infinity at the spiral to tangent point. The radius of the spiral is inversely proportional to its length.

The central angle for the simple curve is chosen from field conditions. The degree of curve is found by consulting tables of design speed prepared by most state highway departments or found in most route surveying books. The station of the PI is found in the field. The length of the spiral is then chosen, depending on the design speed and number of traffic lanes from the appropriate table (see Figures 10-7 and 10-8 for parts of a spiral curve).

The simple curve must be offset by a certain amount given in the tables to allow the insertion of the spiral between the tangent and the curve. This amount is known as the "throw" of the spiral. The length of the circular curve is then reduced by twice the delta of the spiral and the new length of the curve is computed. The calculations for a spiral curve have been greatly simplified by the use of computers. To do a field layout of a spiral curve, set up the instrument on the TS or ST as in a simple curve. Set the points by deflection angles from the tangents. Distances are set by chords

**Figure 10-7** Circular curve with equal spirals

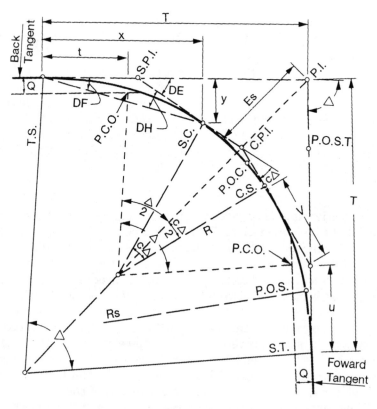

or radial stakeout from the total station. The student is referred to any of the excellent texts on route surveying listed in the bibliography for further study on spirals. However, because of the limited use of spirals in everyday surveying, no more time will be taken here.

## VERTICAL CURVES

In addition to horizontal curves, which turn left and right, roads also have curves that turn up and down, or **vertical curves**. Vertical curves over a hill are called **summit** curves while those at the bottom of the hill are called **sag** curves (see Figure 10-9). The grades between the summits and the sags are tangent to the vertical curves. If these are going uphill, they are plus grades, whereas downhill grades are minus grades. The grade of slope (rise over run) is called the gradient or, more simply, the grade. Do not confuse the term "grade" with other meanings, such as finish grade or subgrade. These are elevations of specific points, whereas "grade" refers to a constant slope. Grades are always computed in the direction of the stationing. Grades may be given as percentage of rise or fall or as a slope that is the rise or fall per foot of run. A +12° slope rises 12 feet in a hundred feet and a +.012° slope rises 1.2 hundredths per foot of run.

## SPIRAL TABLES  a = 2/3

1° in 150 ft.

| Ls' | De | de | df | Q | R + Q | t | x | y | Cs' |
|---|---|---|---|---|---|---|---|---|---|
| Stations | Degrees and Minutes | Deg. & Min. | Deg. & Min. | Feet | Feet | Feet | Feet | Feet | Feet |
| 0.1 | 0°–04′ | 0°–00.20′ | 0°–00.0667′ | 0.000 | 85950.859 | 5.000 | 10.000 | 0.000 | 10.000 |
| 0.2 | 0°–08′ | 0°–00.80′ | 0°–00.2667′ | 0.000 | 42975.107 | 10.000 | 20.000 | 0.001 | 20.000 |
| 0.3 | 0°–12′ | 0°–01.80′ | 0°–00.6000′ | 0.000 | 28650.004 | 15.000 | 29.999 | 0.005 | 30.000 |
| 0.4 | 0°–16′ | 0°–03.20′ | 0°–01.0667′ | 0.003 | 21487.556 | 20.000 | 39.999 | 0.012 | 40.000 |
| 0.5 | 0°–20′ | 0°–05.00′ | 0°–01.6667′ | 0.006 | 17190.023 | 25.000 | 49.999 | 0.024 | 50.000 |
| 0.6 | 0°–24′ | 0°–07.20′ | 0°–02.4000′ | 0.010 | 14325.010 | 30.000 | 59.999 | 0.041 | 60.000 |
| 0.7 | 0°–28′ | 0°–09.80′ | 0°–03.2667′ | 0.016 | 12278.604 | 35.000 | 69.999 | 0.066 | 70.000 |
| 0.8 | 0°–32′ | 0°–12.80′ | 0°–04.2667′ | 0.024 | 10743.780 | 40.000 | 79.999 | 0.099 | 80.000 |
| 0.9 | 0°–36′ | 0°–16.20′ | 0°–05.4000′ | 0.035 | 9550.035 | 45.000 | 89.999 | 0.141 | 90.000 |
| 1.0 | 0°–40′ | 0°–20.00′ | 0°–06.6667′ | 0.048 | 8595.056 | 50.000 | 99.999 | 0.193 | 100.000 |
| 1.1 | 0°–44′ | 0°–24.20′ | 0°–08.0667′ | 0.064 | 7813.703 | 55.000 | 109.999 | 0.258 | 110.000 |
| 1.2 | 0°–48′ | 0°–28.80′ | 0°–09.6000′ | 0.083 | 7162.583 | 60.000 | 119.999 | 0.335 | 120.000 |
| 1.3 | 0°–52′ | 0°–33.80′ | 0°–11.2667′ | 0.106 | 6611.649 | 65.000 | 129.998 | 0.426 | 129.999 |
| 1.4 | 0°–56′ | 0°–39.20′ | 0°–13.0667′ | 0.132 | 6139.419 | 70.000 | 139.998 | 0.532 | 139.999 |
| 1.5 | 1°–00′ | 0°–45.00′ | 0°–15.0000′ | 0.163 | 5730.163 | 75.000 | 149.997 | 0.654 | 149.999 |
| 1.6 | 1°–04′ | 0°–51.20′ | 0°–17.0667′ | 0.198 | 5372.076 | 80.000 | 159.996 | 0.794 | 159.998 |
| 1.7 | 1°–08′ | 0°–57.80′ | 0°–19.2667′ | 0.238 | 5056.121 | 85.000 | 169.995 | 0.953 | 169.998 |
| 1.8 | 1°–12′ | 1°–04.80′ | 0°–21.6000′ | 0.282 | 4775.282 | 89.999 | 179.993 | 1.131 | 179.997 |
| 1.9 | 1°–16′ | 1°–12.20′ | 0°–24.0667′ | 0.332 | 4524.018 | 94.999 | 189.991 | 1.330 | 189.996 |
| 2.0 | 1°–20′ | 1°–20.00′ | 0°–26.6665′ | 0.387 | 4297.888 | 99.999 | 199.989 | 1.551 | 199.995 |
| 2.1 | 1°–24′ | 1°–28.20′ | 0°–29.3998′ | 0.448 | 4093.305 | 104.998 | 209.986 | 1.796 | 209.994 |
| 2.2 | 1°–28′ | 1°–36.80′ | 0°–32.2664′ | 0.516 | 3907.335 | 109.998 | 219.982 | 2.065 | 219.992 |
| 2.3 | 1°–32′ | 1°–45.80′ | 0°–35.2663′ | 0.589 | 3737.546 | 114.997 | 229.978 | 2.360 | 229.990 |
| 2.4 | 1°–36′ | 1°–55.20′ | 0°–38.3996′ | 0.669 | 3581.919 | 119.996 | 239.973 | 2.681 | 239.988 |
| 2.5 | 1°–40′ | 2°–05.00′ | 0°–41.6661′ | 0.757 | 3438.758 | 124.995 | 249.966 | 3.030 | 249.985 |
| 2.6 | 1°–44′ | 2°–15.20′ | 0°–45.0661′ | 0.851 | 3306.620 | 129.994 | 259.959 | 3.409 | 259.982 |
| 2.7 | 1°–48′ | 2°–25.80′ | 0°–48.5992′ | 0.953 | 3184.286 | 134.992 | 269.951 | 3.818 | 269.978 |
| 2.8 | 1°–52′ | 2°–36.80′ | 0°–52.2658′ | 1.063 | 3070.706 | 139.991 | 279.941 | 4.258 | 279.974 |
| 2.9 | 1°–56′ | 2°–48.20′ | 0°–56.0654′ | 1.181 | 2964.974 | 144.989 | 289.930 | 4.730 | 289.969 |
| 3.0 | 2°–00′ | 3°–00.00′ | 0°–59.9986′ | 1.308 | 2866.308 | 149.987 | 299.917 | 5.236 | 299.963 |
| 3.1 | 2°–04′ | 3°–12.20′ | 1°–04.0649′ | 1.443 | 2774.024 | 154.984 | 309.903 | 5.778 | 309.957 |
| 3.2 | 2°–08′ | 3°–24.80′ | 1°–08.2645′ | 1.587 | 2687.524 | 159.982 | 319.886 | 6.355 | 319.950 |
| 3.3 | 2°–12′ | 3°–37.80′ | 1°–12.5975′ | 1.741 | 2606.286 | 164.978 | 329.867 | 6.969 | 329.941 |
| 3.4 | 2°–16′ | 3°–51.20′ | 1°–17.0636′ | 1.904 | 2529.845 | 169.975 | 339.846 | 7.622 | 339.932 |
| 3.5 | 2°–20′ | 4°–05.00′ | 1°–21.6631′ | 2.077 | 2457.791 | 174.971 | 349.822 | 8.314 | 349.921 |
| 3.6 | 2°–24′ | 4°–19.20′ | 1°–26.3957′ | 2.260 | 2389.760 | 179.966 | 359.795 | 9.047 | 359.909 |
| 3.7 | 2°–28′ | 4°–33.80′ | 1°–31.2617′ | 2.454 | 2325.427 | 184.961 | 369.765 | 9.822 | 369.896 |
| 3.8 | 2°–32′ | 4°–48.80′ | 1°–36.2607′ | 2.658 | 2264.500 | 189.956 | 379.731 | 10.639 | 379.881 |
| 3.9 | 2°–36′ | 5°–04.20′ | 1°–41.3931′ | 2.874 | 2206.720 | 194.950 | 389.694 | 11.501 | 389.864 |
| 4.0 | 2°–40′ | 5°–20.00′ | 1°–46.6586′ | 3.100 | 2151.850 | 199.943 | 399.653 | 12.408 | 399.846 |
| 4.1 | 2°–44′ | 5°–36.20′ | 1°–52.0573′ | 3.339 | 2099.680 | 204.935 | 409.607 | 13.361 | 409.826 |
| 4.2 | 2°–48′ | 5°–52.80′ | 1°–57.5892′ | 3.589 | 2050.017 | 209.927 | 419.557 | 14.362 | 419.804 |
| 4.3 | 2°–52′ | 6°–09.80′ | 2°–03.2542′ | 3.851 | 2002.688 | 214.918 | 429.501 | 15.411 | 429.779 |
| 4.4 | 2°–56′ | 6°–27.20′ | 2°–09.0525′ | 4.126 | 1957.535 | 219.907 | 439.441 | 16.511 | 439.752 |
| 4.5 | 3°–00′ | 6°–45.00′ | 2°–14.9837′ | 4.414 | 1914.414 | 224.896 | 449.374 | 17.661 | 449.723 |
| 4.6 | 3°–04′ | 7°–03.20′ | 2°–21.0481′ | 4.714 | 1873.192 | 229.884 | 459.301 | 18.863 | 459.691 |
| 4.7 | 3°–08′ | 7°–21.80′ | 2°–27.2454′ | 5.028 | 1833.751 | 234.871 | 469.222 | 20.118 | 469.655 |
| 4.8 | 3°–12′ | 7°–40.80′ | 2°–33.5760′ | 5.356 | 1795.981 | 239.857 | 479.136 | 21.427 | 479.617 |
| 4.9 | 3°–16′ | 8°–00.20′ | 2°–40.0394′ | 5.697 | 1759.778 | 244.841 | 489.042 | 22.792 | 489.576 |
| 5.0 | 3°–20′ | 8°–20.00′ | 2°–46.6359′ | 6.053 | 1725.053 | 249.824 | 498.940 | 24.214 | 499.531 |
| 5.1 | 3°–24′ | 8°–40.20′ | 2°–53.3655′ | 6.423 | 1691.717 | 254.806 | 508.830 | 25.693 | 509.482 |
| 5.2 | 3°–28′ | 9°–00.80′ | 3°–00.2278′ | 6.808 | 1659.692 | 259.786 | 518.710 | 27.230 | 519.429 |
| 5.3 | 3°–32′ | 9°–21.80′ | 3°–07.2232′ | 7.208 | 1628.906 | 264.764 | 528.581 | 28.828 | 529.372 |
| 5.4 | 3°–36′ | 9°–43.20′ | 3°–14.3513′ | 7.623 | 1599.289 | 269.741 | 538.442 | 30.486 | 539.310 |
| 5.5 | 3°–40′ | 10°–05.00′ | 3°–21.6124′ | 8.054 | 1570.781 | 274.716 | 548.293 | 32.206 | 549.244 |
| 5.6 | 3°–44′ | 10°–27.20′ | 3°–29.0061′ | 8.501 | 1543.322 | 279.690 | 558.131 | 33.990 | 559.173 |
| 5.7 | 3°–48′ | 10°–49.80′ | 3°–36.5327′ | 8.964 | 1516.858 | 284.661 | 567.958 | 35.837 | 569.096 |
| 5.8 | 3°–52′ | 11°–12.80′ | 3°–44.1919′ | 9.443 | 1491.339 | 289.630 | 577.773 | 37.750 | 579.014 |
| 5.9 | 3°–56′ | 11°–36.20′ | 3°–51.9838′ | 9.939 | 1466.718 | 294.597 | 587.574 | 39.729 | 588.926 |
| 6.0 | 4°–00′ | 12°–00.00′ | 3°–59.9084′ | 10.452 | 1442.952 | 299.562 | 597.361 | 41.775 | 598.832 |

**Figure 10-8** Spiral tables. *Courtesy of Washington State Department of Transportation*

**Figure 10-9** Vertical curves

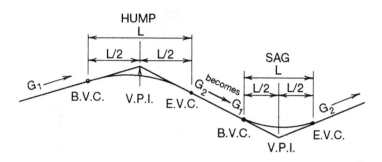

The elevations along the course of the road are determined by the profile, whereas the grades at right angles to the road are cross sections. The elevations are plotted on plan and profile paper, whereas the finished grade of the road is superimposed, calculating for sight distance and stopping distances.

The tangent grades are joined by curves as in horizontal alignment. As in horizontal curves, if a simple curve were put between plus and minus grades, the rapid change in direction would be uncomfortable to the driver. Some of the "Thank you, Ma'ams" (summits and depressions with elevator-like drops) in the southwest desert roads will give you some idea of a simple vertical curve when you become airborne. The parabolic curve of an artillery shell is the best transition between grades. When computing grades, remember that an 8 to 12% grade is considered maximum grade. Heavy trucks and under-powered vehicles will have difficulty with a 12% grade. Between 5 and 8% grade is considered steep for a primary road and maximum for a secondary road.

If the grade is less than 2%, water tends to form "bird baths" on asphalt. You may get by with ½% on smooth finished concrete but not on "broom" finished. Above 12%, liquids tend to run away from solids and the gutter will not be self-cleaning.

The amount of material to be removed or imported is important when planning grades. The earthwork balance must be considered. It is expensive to transport dirt very far, so try to balance the cut and the fill when you can. Earthwork is discussed in Chapter 11.

## THE EQUATION OF A PARABOLIC CURVE

In analytic geometry, the equation of a parabolic curve is given as

$$y = ax^2 + bx + c.$$

The magnitude of $a$ controls the sharpness of the parabola and its sign controls the direction. A positive $a$ is an upward curve or summit curve. A negative $a$ means a downward or sag curve. The property of the parabola is that the tangents to any two points on the curve always intersect exactly halfway between the points of tangency. The change of grade between all pairs of tangents on a parabola is constant, providing the tangents are the same horizontal length.

The length of a vertical curve is always given in stations or horizontal measurement from the points of tangency. If the beginning of vertical curve (BVC) is the origin of the coordinates, the $c$ term drops out of the formula and it becomes $y = ax^2 + bx$. The $x$ coordinate at BVC is zero so that the slope equals $b$, which equals the starting grade or $G_1$. The formula then becomes $y = ax^2 + G_1x$ where $y$ is the vertical distance, and $x$ is the length in stations. $G_1$ is a positive or negative grade in percent and $a$ is in percent units per station.

---

**THE TANGENT OFFSET SOLUTION**

The rule of offsets for $ax^2$ is "Vertical offsets from a tangent to a parabola are proportional to the squares of the distances from the point of tangency." In vertical curves, the distance from the VPI (vertical point of intersection of the tangents $G_1$ and $G_2$) to the MPVC (midpoint of vertical curve) is equal to the distance from the MPVC to the midpoint of a line connecting the BVC and the EVC (end of vertical curve), or distance in elevation BVC-EVC/2 (see Figure 10-10).

The grades diverge by the amount of $G_1$ minus $G_2$, known as the algebraic difference between grades or $G$. The constant rate of change is given by the formula

$$2a = \frac{100(G_2 - G_1)}{L} .$$

Stations on sag curves will have a change in grade of $G_1 + a$ for the first station, $G_1 + 3a$ for the second, $G_1 + 5a$ for the third, etc. The last station grade plus $a$, should equal $G_2$.

A vertical curve is usually stationed on even 50-ft. stations. It is common practice to compute the elevations on the even stations so that horizontal and vertical information for construction will coincide with each other. The BVC, VPI, and EVC are usually set on even stations.

**Figure 10-10** Typical vertical curve

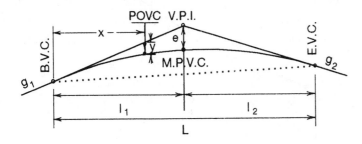

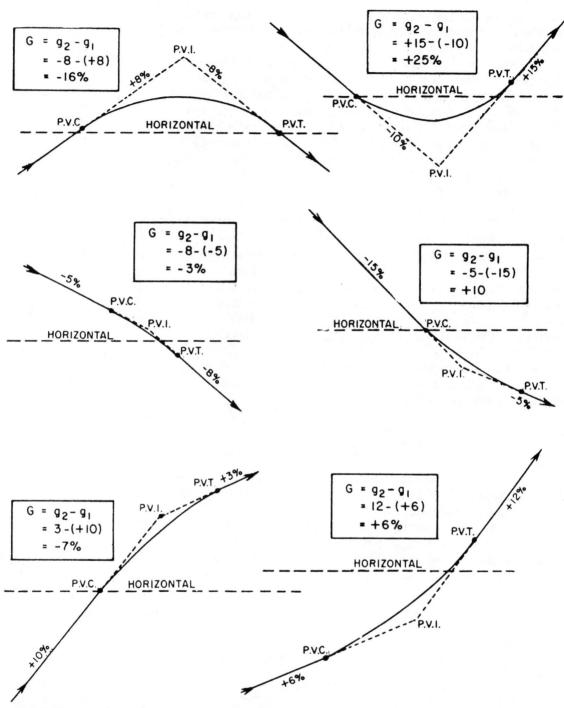

**Figure 10-11** Six types of vertical curves. *Courtesy of U.S. Department of the Army Manual, 1964*

The six possible types of vertical curves and their G calculations are shown in Figure 10-11. A symmetrical vertical curve is the same distance from the BVC to the VPI as it is from the VPI to the EVC, or

$$L_1 = L_2.$$

The best way to learn is by doing, so here are the steps for the tangent offset solution to a summit vertical curve given the following data: $G_1 = +9\%$, $G_2 = -7\%$, $L = 400.00$, VPI station = $30 + 00$, and VPI elevation = $239.12$ ft.

Step 1. Make a table as shown in Figure 10-12. Column 1 is the stations. Column 2 is the elevations on the tangent. Column 3 is the distance from the BVC or EVC, in feet, divided by one half the length, or $x/L_1$ or $x/L_2$. Column 4 is the number in column 3 squared, or $(x/L1)^2$. Column 5 is the vertical offset of the external ($e$) or the distance in feet from the VPI to the MPVC divided into the number in column 4. Column 6 is the number in column 2 minus the number in column 5; this gives the elevation of the POVC (point on vertical curve). Column 7 is the difference between the numbers in column 6; column 8 is the difference between the numbers in column 7. There should not be a great difference between the numbers in column 8 as this is a check on the computations.

**Figure 10-12** Vertical curve table. *Courtesy of U.S. Department of the Army Manual, 1964*

| Stations | Elevations on tangent | $x/l$ | $(x/l)^2$ | Vertical offsets | Grade elevation on curve | First difference | Second difference |
|---|---|---|---|---|---|---|---|
| 28+00 (*PVC*) | 221. 12 | 0 | 0 | 0 | 221. 12 | | |
| | | | | | | +4. 00 | |
| +50 | 225. 62 | ¼ | 1⁄16 | − 0. 50 | 225. 12 | | +1. 00 |
| | | | | | | +3. 00 | |
| 29+00 | 230. 12 | ½ | ¼ | − 2. 00 | 228. 12 | | +1. 00 |
| | | | | | | +2. 00 | |
| +50 | 234. 62 | ¾ | 9⁄16 | − 4. 50 | 230. 12 | | +1. 00 |
| | | | | | | +1. 00 | |
| 30+00 (*P. V. I.*) | 239. 12 | 1 | 1 | − 8. 00 | 231. 12 | | +1. 00 |
| | | | | | | . 00 | |
| +50 | 235. 62 | ¾ | 9⁄16 | − 4. 50 | 231. 12 | | +1. 00 |
| | | | | | | −1. 00 | |
| 31+00 | 232. 12 | ½ | ¼ | − 2. 00 | 230. 12 | | +1. 00 |
| | | | | | | −2. 00 | |
| +50 | 228. 62 | ¼ | 1⁄16 | − . 50 | 228. 12 | | +1. 00 |
| | | | | | | −3. 00 | |
| 32+00 (*PVT*) | 225. 12 | 0 | 0 | 0 | 225. 12 | | |

Step 2. Compute the stations and elevations for the BVC and the EVC from the station for the VPI. The station for the BVC is

$$30 + 00 - 200.00 = 28 + 00$$

and the station for the EVC is

$$30 + 00 + 200 = 32.00.$$

The elevation of the BVC is .09 (9%) times $L_1$ or 200 ft.

$$.09 \times 200 = 18 \text{ ft.}$$ subtracted from the elevation of the VPI or $239.12 - 18 = 221.12$.

Therefore the elevation of the BVC is 221.12.

The elevation of the EVC is found in the same way. The $G_2$ grade is $-7\%$ and the $L_2$ length is 200 ft. The calculation is

$$.07 \times 200 = 14 \text{ ft.}$$

The elevation of the EVC is

$$239.12 - 14 = 225.12.$$

Step 3. Calculate the elevations for each 50-ft. station next. For $G_1$ the 50-ft. tangent grades are .09 times 50 or 4.50 ft. per 50. The BVC elevation is

$$221.12 + 4.45 \text{ to } 28 + 50 = 225.62.$$

Enter this number in the second column opposite $28 + 50$. For $29 + 00$ add 4.45 to the elevation of $28 + 50$ and get 230.12, plus 4.50 is 234.62 for $29 + 50$, plus 4.50 gives 239.12 for $30 + 00$. This checks with the given elevation for the VPI.

The calculations for $G_2$ are done the same way. Multiply .07 by 50 and get 3.5 ft. of difference per 50 ft. of distance. The VPI elevation is 239.12 minus 3.5 for an elevation of 235.62 at $30 + 50$, $235.62 - 3.50$ equals 232.12 at $31 + 00$, $232.12 - 3.50 = 228.62$ at $31 + 50$, $228.62 - 3.50$ equals 225.12. This is the calculated elevation at the EVC station 32.00.

Step 4. Calculate $(e)$ the middle vertical offset at the VPI. First, find $(G)$, the algebraic difference of the grades using

$$G = G_2 - G_1 \text{ or } -7 -(+9) = -16\%.$$

A common formula for finding the middle vertical offset is

$$e = \frac{LG}{8},$$

$L$ is the length in stations, 8 constant, and $G$ is as calculated before.

$$e = \frac{4\,(-16)}{8}, \; -8.00 \text{ ft.}$$

The negative sign indicates that $e$ is subtracted from the elevation of the VPI.

Step 5. Compute the vertical offsets at each 50-ft. station using the formula

$$y = [x/(\tfrac{1}{2}L)]^2 e.$$

To find the vertical offset at any point on a vertical, find the

$$\frac{x}{\text{ratio} \quad \tfrac{1}{2}L,}$$

square that number and multiply by $e$.

For example, at station 28 + 50, the ratio is

$$50/200 = .25, .25^2 = .0625.$$

The vertical offset at 28 + 50 equals

$$.0625 \text{ times } -8 \text{ equals } -0.50.$$

Repeat the calculation for each of the remaining stations.

Step 6. Compute the curve grade elevations for each of the 50-ft. stations. When the curve is on a crest, the sign of the offset is negative. Subtract the vertical offset (the number in column 5) from the tangent grade (column 2). For example, the tangent grade at 28 + 50 is 225.62, minus .050 is 225.12, the elevation for 28 + 50 on the vertical curve. Compute the curve-grade elevations for each of the remaining stations in a similar manner and enter the grades in column 6.

Step 7. Locating the turning point on a vertical curve is necessary for finding the crest on a summit vertical curve and, more importantly, for finding the low point on a sag vertical curve. The low point is the place to install the drainage structure for the curve. Also, if the curve is under an overpass, the turning point is the point of maximum clearance. Only when both tangents have the same rate of grade ($G_1 = G_2$) will the turning point be directly over or under the VPI. Otherwise, the turning point will be on the same side of the curve as the tangent with the least percent of slope. The horizontal distance to the turning point of the curve is measured from the point of tangency of the tangent with the least slope. The horizontal distance is found by using the formula:

$$x_t = G$$

where $x_t$ is the distance from the BVC or EVC in feet. The lesser slope is $g$ (ignoring the sign); $L$ is the length of the curve in stations and $G$ is the algebraic difference in the slope ($G_2 - G_1$). On the example curve the computation is

$$x_t = \frac{7(4)}{16} = 1.75.$$

The turning point is 1.75 stations from the EVC or 32 + 00 minus 175 equals 30 + 25. The vertical offset at 30 + 25 would be

$$\left( \frac{1.75}{2} \right)^2$$

times 8 equals 6.12 above the tangent grade of 1.75 times 7 plus 225.12 = 237.37. The tangent grade of 237.37 minus the vertical offset of 6.12 gives a grade of 231.25 at station 30 + 25 for the high point of the vertical curve.

Step 8. Always be sure to check your work. One of the characteristics of a parabolic symmetrical curve is that the second differences between successive curve elevations at full stations are a constant. In computing the first and second differences (columns 7 and 8) be sure to pay attention to the sign. There may be a slight difference due to rounding error but not enough to cause any problem in the final calculation. Do not worry about small differences in the second difference. They do not always come out even as in this curve. This was how to do a vertical curve by tangent offsets. The other method uses the parabolic formula to solve the curve elevations.

To solve a vertical curve by equation, the following formulas will be used:

$$r = \frac{G_2 - G_1}{L}$$

and

$$y = \frac{r}{2} x^2 + G_1 x + c$$

$r$ being the rate of change and $c$ being the elevation of the BVC. You are familiar with all other terms. For this curve: $G_1 = -6\%$, $G_2 = +2\%$, VPI station $= 20 + 00$, $L = 400.00$ ft., and the elevation $c$ at the BVC $= 348.52$. Subtracting the first formula, we get:

$$r = \frac{+2-(-6)}{4} = +2.$$

Subtracting in the second formula we get:

$$y = \frac{+2}{2}(.5^2) + (-6).5 + 348.52$$

for the first 50-ft. station. Solving for $y$ we get

$$y = 1(.25) + (-3) + 348.52.$$

Combining terms we get

$$y = -2.75 + 384.52 = 345.77$$

for the elevation at $18 + 50$. For $19 + 00$,

$$y = \frac{2}{2}(1^2) + (-6)1 + 348.52.$$

Solving we get

$$y = 1(1) + (-6) + 348.52.$$

Combining we get

$$y = -6 + 348.52 \text{ or } y = 342.52.$$

At $19 + 00$ the curve elevation is 342.52. For $19 + 50$

$$y = \frac{+2}{2} (1.5^2) + (-6)\, 1.5 + 348.52,$$

solving $y = 2.25 + (-9) + 348.52,$
combining $y = -6.75 + 348.52$, or
$$y = 341.77.$$

Continue through the rest of the stations using the same method and be sure to check into the calculated grade for the EVC. The vertical curve by equation is preferred by many because it is easy to program into a programmable calculator.

**UNEQUAL TANGENT VERTICAL CURVES**

Unequal tangent vertical curves are used when an equal tangent vertical curve will not fit the contour of the ground. An unequal tangent vertical curve is similar to a compound horizontal curve in that it is really two vertical curves joined at the point of a compound vertical curvature (PCVC). The PCVC is vertically above or below the VPI of the unequal tangent curve. A line drawn from the midpoint of $G_1$ to the midpoint of $G_2$ will pass through the PCVC and the grade of that line can be found using the formula

$$\frac{G_1\, L_1 + G_2\, L_2}{L_1 + L_2}$$

(see Figure 10-13). At the PCVC the $r$ value of the first curve changes to the $r$ value of the second curve. Solve the unequal tangent vertical curve as two equal tangent vertical curves having the respective lengths of $L_1$ and $L_2$ of the unequal tangent curve.

**Figure 10-13** Unsymmetrical vertical curve

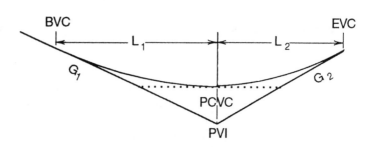

# Review Questions

1. Name four types of horizontal curves.
2. Define the following horizontal curve elements:
   A. PI
   B. I or Delta
   C. R
   D. POC
   E. PC
   F. PT
   G. L
   H. T
   I. LC
   J. E
   K. M
   L. D
3. How far apart should the stakes be set on a 5° curve?
4. What are the deflection angles used when laying out a curve?
5. Give the formula for finding the LC of a curve.
6. Calculate the deflection angles and short and long chords to lay out a horizontal curve.
   Given I = 35°, R = 250 ft., PI = 14 + 25
   Stakes are to be set on half and full stations.
7. Define a transitional spiral curve.
8. Calculate the elevations for a vertical curve.
   BVC = 4 + 25, $G_1 = -4.5\%$, $G_2 = -7\%$, VPI elev. = 854.20, $L = 600$ ft. Stakes are to be set on half stations.
9. Calculate the elevations for a vertical curve.
   VPI sta. = 14 + 75.50, $G_1 = -4.5\%$, $G_2 = +3\%$, VPI el = 145.72, $L = 400$ ft., stakes at half and full stations.
10. Calculate the deflection angles and chords to lay out a horizontal curve, to the right, with a radius of 250.00 ft. and a central angle of 472° 35'. The PI station is 29 + 75.20. Stakes are to be set on 35-ft. offsets. Each side of the center line on stations and half stations.

# Construction Surveys

## Objectives

After completing this chapter and the field practice, the student should be able to:

1. State why not discussing the survey with nonproject people is so important, and to whom to refer any questioner.
2. Relate the types of surveys needed for a common construction project.
3. Set and mark a grade stake with the proper offset and cut or fill to grade.
4. Given a grade stake with the cut or fill to finished grade, find the daylight or catch point for a 1.5 : 1 and 2 : 1 slope.
5. Given a building plan, set the offset stakes for the corners and check for square by measuring the diagonals.
6. Set sewer offset stakes and mark the cuts on the stakes.
7. Set curb offset stakes for a street and stake the curb returns for a 90° intersection using deltas or stationing.
8. Given the cuts and fills for the cross sections of a road or borrow pit, calculate the earthwork volume in cubic yards.
9. Give the hand signals for stop and emergency stop used on construction projects.
10. Recognize the safety hazards to watch for on a construction site and suggest methods to avoid being injured by them.

The majority of surveying work is construction surveying of one kind or another. Most of the surveyors the public sees are working along the streets doing lot surveys; off the public roads, many more are making surveys related to building one thing or another. Construction staking is a specialty type of survey and many survey technicians do only construction surveys. Most construction surveys are made by a firm of consulting engineers made up of licensed surveyors plus various types of engineers and architects. These firms provide a complete service to anyone wanting to build something on the ground, from topography to **"as-built"** surveys of the completed projects.

## THE PRELIMINARY OR ROUTE SURVEY

The first survey made on most construction projects is the preliminary or route survey. On large area projects such as subdivisions, dams, commercial developments, or recreational sites, the topographical survey will probably be made by aerial photomapping. The control for the aerial photography will usually also be used for project control.

On long projects such as highways or pipelines, the route will usually be picked out roughly, using U.S. quadrangle maps. Then a preliminary route survey will be made on the ground. The control points for the preliminary survey will also be the control points for construction. With the use of GPS receivers, the work of preliminary surveys has been greatly simplified. What used to take months of traversing and line brushing can now be done in weeks by fewer technicians using all-terrain vehicles and helicopters.

## BOUNDARY SURVEYS

Once the final site or route has been chosen, legal surveys of the boundaries must be made by a registered land surveyor. When surveying property for a construction project, the reason for the survey should not be discussed with anyone other than project personnel. The reason for this is that the negotiations for right-of-ways or purchase may not have been completed by the principals and they may not want the word to get out until they are ready to announce it to the public. If you are asked what the survey is for or who is having it done, refer the person asking about it to the party chief. Do not be rude or discourteous to the inquirer. Remember, you are your employer's representative in the field.

## PROJECT CONTROL POINTS

Once the project boundaries have been established, the preliminary construction plans are prepared and the project control points are established and referenced. A local grid is usually established for small projects, using coordinates chosen so that negative coordinates will not be used. **State plane coordinates** may be used for larger projects to tie the project into other surveys in the area. Some states require that at least one corner be tied into the state plane system.

If the project is covered with vegetation, the next step will be setting limits for clearing and grubbing. Topsoil may be removed and stockpiled on or near the job site. If trees are to be left, be sure they are clearly marked to save. If the project is large enough, a temporary office may be set up on site for a project superintendent and the storage of stakes and lath. All work done on the site will be coordinated through the construction superintendent.

## LIMITS FOR CLEARING AND GRUBBING

The next step will be setting rough grade stakes for the equipment. Roads and building sites are staked so that the rough cut and fill work can be done by the grading contractor. Stakes for rough grade are usually set on an offset determined by the grading contractor. Five-foot offsets for roads are common. Rough grades are normally run to the nearest tenth of a foot and are not tacked. A hub is driven into the ground and a lath is put in a couple of tenths away on the side away from the work (see Figure 11-1). This allows the grade checker to transfer the grade to the work area without needing to move the lath. The offset is marked on the face of the lath facing the work, and the cut or fill is marked below the offset. The station, or description of the point, is marked on the back of the lath.

## ROUGH GRADE STAKES

If the cuts or fills are more than a couple of feet, the face will be sloped to prevent caving or sliding of the material. Slope stakes are set at the top of cuts and at the toe, or bottom, of fills. The amount of slope needed for **repose** (stability) of a slope will be determined by a soils engineer.

Most cuts in normal ground are a slope of 1.5 to 1 or, if the soil is firm such as decomposed granite, 1 to 1. These are referring to the amount of run, or horizontal distance to rise, or difference in elevation, respectively. A 1.5 to 1 slope means that for every foot of cut or fill, the slope has 1.5 ft. of horizontal run (see Figure 11-2).

Fills require the placement of loose material in lifts or layers as the slope rises. Each of these lifts must be compacted and tested. As each lift is approved by the soils technician, another is laid on top of the previous one but set back by the amount necessary to maintain the slope. The grade is usually checked by a grade-checker using a hand level. Most fill slopes are 2 : 1. The face is compacted and graded smooth as the fill progresses.

For high slopes, a step (or bench) is built into the face every so many vertical feet. These benches allow water running down the slope to drain off across the face into a collection ditch and be carried away. This prevents erosion of the face of the slope. After the slope is built, the surveyors will usually be called back to check

## SLOPE STAKES

**Figure 11-1** Slope stakes

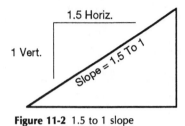

**Figure 11-2** 1.5 to 1 slope

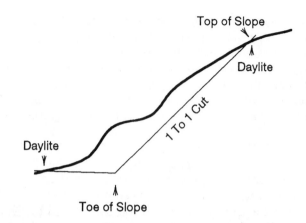

**Figure 11-3 (Right)** Daylight points
on a cut slope

the location of the top or toe and to set new grade stakes for whatever is to be built near the finished slope.

To set slope stakes, take a shot at the location of the top, or toe. Compare the elevation of the shot with the finished elevation of the point. Then move the rod the calculated horizontal distance for the cut or fill and take another reading. Repeat this process until the location and elevation match the calculated cut or fill. This point is known as the **catch point**, or **daylight point**, of the slope (see Figure 11-3). Set a stake at this point and mark the cut or fill on the lath, or guard, stake. At a convenient distance away from the top, or toe, place a reference stake. Mark the offset and difference in elevation to the top, or toe, stake on the lath as well as the information from the referenced stake (see Figure 11-4).

Some programs for pocket calculators have a slope-staking routine that greatly simplifies the necessary calculations. With practice, the survey technician will be able to anticipate the location

**Figure 11-4** Slope stakes

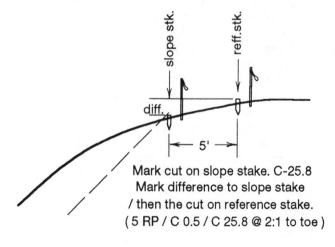

Mark cut on slope stake. C-25.8
Mark difference to slope stake
/ then the cut on reference stake.
( 5 RP / C 0.5 / C 25.8 @ 2:1 to toe )

of the trial catch point and need only a couple of trial points to get the correct location. As you move out from one point to the next one, estimate the amount of rise or fall of the surface of the ground. Figure in your head how much horizontal distance is required to the catch point. Sometimes, you can use the rod and your hand level to measure the rise and run and get pretty close on the first try.

Some total stations have slope-staking programs that allow the setting of stakes on lines other than the line on which the instrument is set up. After the rough grading is done, the grading contractor will ask for subgrade stakes to be set.

---

## GRADES FOR BUILDINGS

The grades for buildings are usually set on offsets to the building corners. Each corner will require two stakes, one on each building face projected out a specified distance from the building (see Figure 11-5). Show two grades on each stake: one for subgrade and one for finished floor elevation. Mark these grades in feet and inches; carpenters do not like tenths and hundredths of feet. For inside corners, a stake on the opposite side of the building is necessary for line.

For large buildings, piles are driven in the ground to support the weight of a tall structure. If piles are needed, the contractor will want stakes set for each pile. The offsets are set out from the construction area on what are called rows and bents (see Figure 11-6).

When piles are being driven, an instrument is set up on each of the corresponding offset stakes. Horizontal and vertical alignment of the pile is given to the pile-driver operator by radio. After the piles are driven, the superintendent will ask for cutoffs to be marked on the piles. Be sure to turn through these grades, as a mistake can be very expensive.

**Figure 11-5 (Below left)** Building corner staking

**Figure 11-6 (Below)** Piling control layout

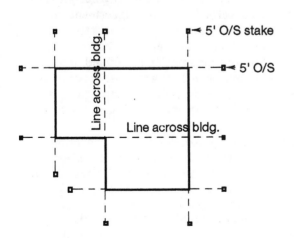

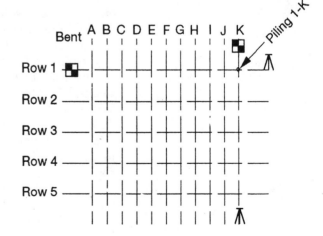

## UNDERGROUND UTILITIES

All the underground utilities should be installed before the streets are subgraded. This may include sanitary sewer, storm sewer, water, gas, telephone, cable TV, and electricity. Run the center lines of each utility first and set stakes at each angle point. The offset stakes are set at the distance decided on by the utility contractor. This will depend on the width of the equipment to be used to dig the trench. The side of the trench where the stakes are to be set will depend on which way the dirt is removed from the trench and where it will be dumped.

## DRAINAGE SYSTEMS

Drainage systems are of two types, sanitary and storm. The principal is the same, it always runs downhill. Downhill, in this case, is a grade of more than 2% and less than 12%. The grade will be calculated by the licensed sanitary engineer to assure gravity flow through the pipe. If the grade is too shallow, a pressure line is installed with pumps to force the flow through the pipe.

A word of caution! NEVER TAMPER WITH THE SCREWS HOLDING THE MANHOLE COVER ON A SEWER MANHOLE. Pressure sewer manhole covers are screwed on because the pressure is sometimes very high inside. If the screws are loosened, whatever is under pressure inside comes spraying out the side and it is nearly impossible to get the lid back on once the seal is broken. Some sewer manhole lids are of the locking type that must be turned to open but these are not normally under pressure. Most manhole lids can be safely removed with a pick or manhole hook. Be careful for they are very heavy.

## SANITARY SEWER SYSTEMS

Sanitary sewer systems begin at the treatment plant and spread out like branches on a tree. The main trunk lines are designed to carry all the flow from the collector lines to the treatment plant. The collector lines carry the flow from the street sewers that carry the flow from the house connections. The engineer bases the size of the pipe on the average daily consumption of 100 gal. for each person in the house. This is determined by the number of bedrooms planned for the home. This number is multiplied by a peak flow factor to take care of the morning rush and half time at the Super Bowl.

The house connection enters the street sewer at a "Y" connection in the pipe. Sanitary sewer plans will usually give a station for the house connection and the Y. The manholes and Ys are staked on the center line and then offset in the proper direction. A 6-ft. offset is standard for most sanitary sewers. The standard manhole is 4-ft. wide so the inlet and outlet stakes must be set 2-ft. from the center of the manhole. A manhole must be set at every change in direction, horizontal or vertical, in the sewer line. The exceptions are the Ys and clean outs. Manholes are also required every 600 ft. on straight runs.

Some sewer contractors are using lasers to set the pipe and will require cuts only at the manholes. Others will require offset stakes every 25 to 50 ft. with cuts to flow line marked on each stake.

A grade sheet is usually provided to the contractor at the end of the staking or each day on a large job. A stake is set on each house connection and an offset stake is set at the property line with the cut to the flow line of the pipe. After the pipe has been laid, the surveyor may be called back to perform an "as-built" survey of the complete sewer.

---

**STORM SEWERS**

Storm sewers begin at the natural water drainage carry-off features in the drainage area; these may be rivers or lakes. If no natural drainage is left due to building, a storm channel is used. As with a sanitary sewer, a series of collector pipes or structures fan out upstream. This time, the upper end of the system is the ridge of the mountains surrounding the area where the water begins its flow to the outlet drainage or the water body that collects all the flow in the watershed.

As we build more and more paved areas, the natural drainage and **percolation** of the water into the ground are changed. Storm drains must be designed to carry the runoff into channels where it will flow off and not cause flooding. There are intermittent periods of increased precipitation known as the 10-, 50-, and 100-year floods. The increased flows during these periods cause extra problems for planners. Some flood-control districts do not allow any building within the 100-year flood plain.

An artificial drainage system consists of ditches, either paved or unpaved. When a ditch passes under a structure it becomes a **culvert**. There are two types of culverts: pipe culverts and box culverts. Pipe culverts may be arched or round. Box culverts may be square or rectangular. Both are constructed of galvanized iron or reinforced concrete. The ends have head walls and water is directed into the opening by wing walls.

The underground drainage system or storm sewer is usually buried pipe called the trunk or main line. It includes a system of collector lines fed by catch basins or inlets. The inlet built into a concrete curb is called a curb inlet. This feeds water into the catch basin, which is connected to the storm sewer by a pipe. An inlet that is in the gutter, or flow line, is covered with a grate that should be self-cleaning. These inlets are called **appurtenances**. Other appurtenances are the manholes and junction boxes.

As with the sanitary sewers, manholes are installed at all points where the storm sewer changes direction or grade. Distances between manholes should not be more than 500 feet. When measuring down from the manhole rim to the pipes, the flow line of the pipe is called the invert. The inside top of the pipe is called the obvert. Normally, the drop across a 4-foot manhole, invert to invert is .10 of a foot.

The quantity of water that will flow through a pipe is controlled by the size of the pipe, the grade, and the type of pipe used. Water will flow better through a smooth-walled pipe than a rough-walled pipe such as corrugated metal. A large pipe will carry more water than a small one. A steep grade will cause the water to flow faster and carry more water than a shallow grade. A smooth, concrete pipe will have less friction between the wall and the water. All of these factors control the size, grade, and type of pipe selected by the designer.

## STATIONING

When referring to either type of sewer, stationing always begins at the start or lower end and increases upstream. As you stand at the point of lower stationing and face upstream, the left side of the pipe is on the left and the right side of the pipe is on the right.

## STREETS OR ROADS

Streets or roads in urban and suburban areas are usually paved, either with asphalt or concrete. They may or may not have curbs and gutters. The cross fall of a street is how much the pavement drops from the center line to the flow line or edge of pavement. The cross fall is usually 2% for paved streets. The parkway is that part of the street lying between the back of curb or edge of pavement and the property line. It should also have a 2% fall away from the property-line grade to the top of curb or flow line of the drainage along the edge of the road.

Street plans will have the plan view on the upper portion. A profile of the street on the exaggerated scale on the ruled section is at the bottom. A detail sheet will have the cross section and the details for any appurtenances to be constructed within the right of way. The elevations are for finished surface. You must subtract the thickness of the paving and the ABC or subbase from the finish grades to get the subgrade elevations (see Figure 11-7).

**Figure 11-7** Typical street cross section

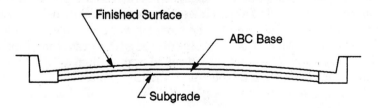

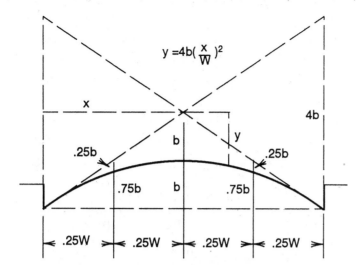

$$y = 4b \left( \frac{x}{W} \right)^2$$

**Figure 11-8** Street cross section

The cross section of the street shows a straight 2% cross fall. Actually the cross section is in the form of a parabola. You will need to set grades for the ¼ points. Figure 11-8 shows the cross section and the formula for calculating the ¼ points or you may use your vertical curve program.

**OFFSET STAKES**

The center line of the street is generally laid out first along the tangents. The BC and EC points are then staked. The offset stakes will usually be set at a 5-ft. offset (o/s) to back of curb, if there is a curb. The o/s stakes will be 5-ft. off the edge of paving, if there is to be a rolled edge on the pavement. This is done because the curbs are most often poured first. Then the subgrade for the pavement is placed between the curbs. Yellow-topped 1 × 2s are used for grading the surface to fine grade (see Figure 11-9).

After the tangents are poured for the curb, the returns, or curved portions, are poured. The layout of the curb returns is usually set on even lengths of arc or deltas (see Figure 11-10). For a 90° curb return you will normally set the BCR (begin curb return) in the direction of traffic flow. Then set ¼ Δ, ½ Δ, ¾ Δ, and the ECR (end curb return).

If the street is to have concrete cross gutters, you will set an "X" point at the intersection of the flow lines (see Figure 11-11). The X is called the **spandrel point** and is usually given on the street plans.

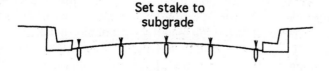

Set stake to
subgrade

**Figure 11-9** Finish-grade stakes for street

**Figure 11-10** Staking curb return

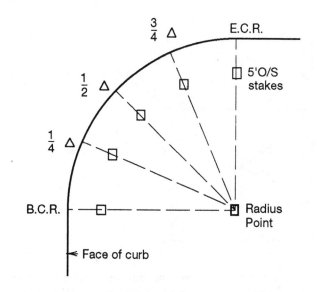

**Figure 11-11** Spandrel points

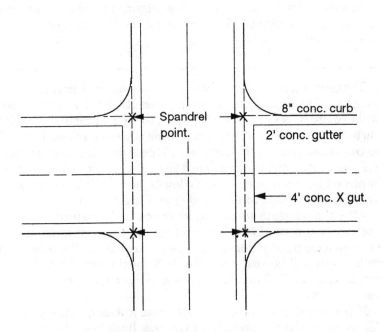

**"BLUE TOPS"**

After the subbase is laid and compacted, the survey crew will come back to set "blue tops" for paving. Blue tops are 1×2 stakes set to grade for the top of the ABC and marked with blue kiel on the top. The asphalt paving is laid on top of this grade. Blue tops are usually set on center line and the ¼ points. When the street has been paved, the sidewalks are poured and the curb is backfilled.

In the office, after the center-line profiles and cross sections have been plotted, the earthwork quantities must be calculated. The cross section of the subgrade is plotted to scale over the cross section of the original ground and the coordinates of the grade breaks are computed. The end areas are calculated and cuts and fills are separated by area. The areas of two adjacent stations are averaged and the result is multiplied by the distance between stations. This gives the approximate volume in cubic feet. Cubic feet are divided by 27 to get cubic yards. This method is called the "average end area" and is used for fairly regular terrain. For more precise volumes, the **prismoidal formula** is used. The formula for average end area is:

$$V = \frac{A_1 + A_2}{2} \; L.$$

The prismoidal formula is

$$V = \frac{A_1 + 4A_m + A_2}{6} \; L$$

where $L$ is the distance from $A_1$ to $A_2$. Note that $A_m$ is not the average of $A_1$, $A_2$ but is computed using the average of the corresponding distances on $A_1$ and $A_2$.

**EARTHWORK QUANTITIES AND END AREAS**

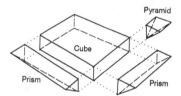

**Figure 11-12** Borrow pit excavation shapes

**BORROW PITS OR RAISED PADS**

On regular shapes such as borrow pits or raised pads it is easier to break the total volume down into regular geometric figures. Geometric formulas for cubes, prisms, and cones can be used to find the volumes of the several parts, which are then added together for the total volume (see Figure 11-12).

**COMPUTER SURVEY PROGRAMS**

Most computer survey programs have earthwork subprograms. The Earthwork/ROADS Volume 3 of the ESP 200 Engineering Survey Package by PacSoft of Kirkland, Washington, has the ability to take data directly from field notes to calculate and plot profiles and cross sections. New cross sections can be interpolated between existing sections, if necessary. Roadway templates can be created and stored. The templates can be used for "branching." That is, they can be used to make decisions about design changes, depending on whether the section meets the template in cut or fill. It can help decide whether to put in a ditch, if the decision point is in cut, or a simple slope, if the cut falls in fill. An "unless/then" type of decision can be to use a 2 : 1 slope unless the catch point falls outside the right of way. In that case, use whatever slope is

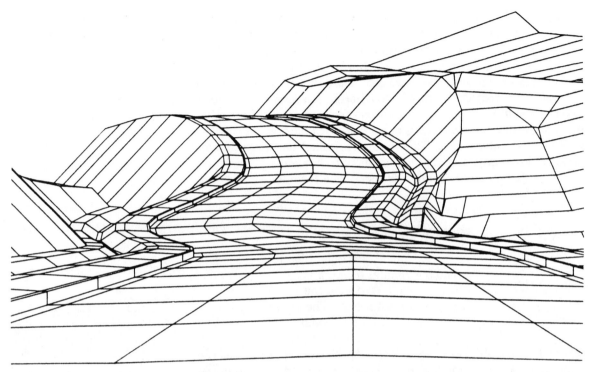

**Figure 11-13** A drive-through perspective. *Courtesy of PacSoft*

necessary to put the catch point inside the right of way and print a warning message.

Layers can be used on volumes to handle cut and fill, subbase, asphalt, concrete, or any other case where a separate type of material volume is needed. A choice of average-end-area or prismoidal methods of calculations is offered. The printout includes area of cut, of fill, bank volumes, loose volumes, and mass ordinate for each station or cumulative. An interesting and useful computer feature program is the "drive-through" **perspective** that enables the designer and the client to see approximately what the finished road will be like and make any changes before "the concrete is poured" (see Figure 11-13).

When calculating cuts and fills on a grading project, be sure to keep in mind the cost of moving dirt. The cuts and fills should balance over as short a haul as possible under existing conditions.

## A WORD OF CAUTION

A word of caution, WHENEVER YOU ARE WORKING AROUND HEAVY EQUIPMENT BE CERTAIN THAT ALL THE OPERATORS KNOW THAT YOU ARE ON THE JOB. Remember that the operators have their hands full with what they are doing. They will not run over you deliberately, but often their attention is on something besides where you are standing. You will be required to wear a hard hat and a red vest for protection and visibility on all construction jobs. Equipment must not be left unattended around any job site. Be especially careful of tapes stretched out along the ground, for they are nearly impossible to see. To a heavy-equipment operator, the universal signal for stop is a clenched fist held at arm's length above your head. Both fists held up at the same time means "emergency stop." Do not use this signal unless it is really an emergency. The operators will stop at all costs and may injure themselves or damage the equipment in the process.

As a construction surveyor, you may have the opportunity to work in international construction. The skills you have are your ticket to work in other countries on major construction projects. If you do plan to work overseas, it is a good idea to be familiar with the metric system that is used in most countries of the world. It would also benefit you to be familiar with the language spoken in the country where you will be working. The libraries in most big cities will have a section of telephone books for most major cities in the United States. Look in the yellow pages, under Construction Companies, for the addresses of those companies that do international construction, then send them a resume. You will also hear of jobs available overseas from the people you work with. The pay is good, but the working conditions can be difficult and you may need to sign a one-year contract. Housing is usually furnished and the food is good. If you quit, however, it may be necessary to pay your own way back to the United States.

**INTERNATIONAL CONSTRUCTION**

## Review Questions

1. Should you discuss the project you are working on with persons other than those involved in the project?
2. What should you say to someone who asks what you are doing?
3. How accurate should you mark rough grades for cut and fill?
4. On what side of the cut/fill stake should the guard stake or lath be set?
5. When marking the grade stake lath, what should be shown and on which side should the information be shown?
6. What does 1.5 : 1 mean on a slope stake?
7. You are setting slope stakes and the party chief tells you to go out 10 feet more horizontally. What should you do while moving out the 10 feet?
8. How do you determine on which side of the sewer center line to set the offset stakes?

9. Should you remove the screws from a sewer manhole to check the depth?

10. In what direction does the stationing increase on a sewer line?

11. What side is the right side in stationing?

12. What is the purpose of a blue top stake on a paving job?

13. The following field notes are for 200 feet of profile and cross sections for a road. The road will be 25-feet wide on each side of the center line for a total of 50 feet of roadway. The finished center line grade at station 1+00 is 84.00 feet. The road has a +2% grade increasing up station. Fill slopes for the road will be 2 : 1 and cut slopes will be 1.5 : 1. The road has a 2% cross fall. Make a plan and profile for the road and calculate the total volumes of cut and fill.

| Sta. | 1+00 | 50'Lt | Elev. | 82.4 |
|------|------|-------|-------|------|
|      | "    | 25'Lt | "     | 83.0 |
|      | "    | CL −  | "     | 84.2 |
|      | "    | 25'Rt | "     | 85.6 |
|      | "    | 50'Rt | "     | 86.9 |
|      | 1+50 | 50'Lt | "     | 86.0 |
|      | "    | 25'Lt | "     | 86.5 |
|      | "    | CL −  | "     | 87.6 |
|      | "    | 25'Rt | "     | 88.8 |
|      | "    | 50'Rt | "     | 90.3 |
|      | 2+00 | 50'Lt | "     | 89.5 |
|      | "    | 25'Lt | "     | 90.0 |
|      | "    | CL −  | "     | 91.0 |
|      | "    | 25'Rt | "     | 92.0 |
|      | "    | 50'Rt | " .   | 93.6 |
|      | 2+50 | 50'Lt | "     | 93.0 |
|      | "    | 25'Lt | "     | 93.5 |
|      | "    | CL −  | "     | 94.3 |
|      | "    | 25'Rt | "     | 95.2 |
|      | "    | 50'Rt | "     | 96.9 |
|      | 3+00 | 50'Lt | "     | 96.5 |
|      | "    | 25'Lt | "     | 97.0 |
|      | "    | CL −  | "     | 97.6 |
|      | "    | 25'Rt | "     | 98.4 |
|      | "    | 50'Rt | "     | 100.2 |

# Public Land Surveys

**12**

## Objectives

After completing this chapter, the student should be able to:

1. Apply the general rules for the survey of public lands to retracement surveys.
2. Distinguish between township and range lines.
3. Given a township map, identify the sections and government lots.
4. Given a section map, identify the regular aliquot parts of the section.
5. Define existent, obliterated, and lost corners.
6. Describe the proper method of searching for a corner.
7. Apply the proper single or double proportionate method in restaking a lost section corner:
   A. interior;
   B. exterior or township corner.
8. Compare the tangent and secant methods for setting township lines or parallels.
9. Define meander corner and closing corner and give the proper method to reset each one, if lost.
10. Prepare a section map showing the aliquot parts and the regular dimensions and acreage of each part.
11. State the controlling principal meridian and base line for your area.
12. Identify "hacks" and "blazes" on trees and state why they should not be used any longer.
13. Describe a standard monument:
    A. in old surveys;
    B. in modern surveys (1973 manual).
14. Define "corner accessories" and give examples of each type.

15. Justify the use of aerial photogrammetry and analytical phototriangulation in the re-retracement surveys of public lands.

16. Recognize and cite evidence for the need to use astronomical azimuth in modern surveys.

## THE MEETING ON THE NORTH BANK OF THE OHIO RIVER

Be it ordained by the United States in Congress assembled, that the territory ceded by individual states to the United States, which has been purchased of the Indian inhabitants, shall be disposed of in the following manner.

This is the way the surveys of the public lands of America began. Although a surveyor was appointed from each of the states, only eight met geographer Thomas Hutchins on the north bank of the Ohio River that frosty morning in 1785. Edward Dowse from New Hampshire, Benjamin Tupper of Massachusetts, Isaac Sherman from Connecticut, and Absalom Martin of New Jersey were present. Also present were William Morris from New York, Alexander Parker from Virginia, James Simpson from Maryland, and Robert Johnston of Georgia.

On September 30, 1785, Hutchins and his party started from a wooden post, set by Andrew Ellicott, at the high water mark on the west boundary of Pennsylvania. By October 8, 1785, the party composed of Hutchins, the eight surveyors, and about thirty chainmen and axmen, had run about four miles of line to the west of the point of beginning. The line was run using a compass and a two-pole Gunter's chain. On October 8, work was stopped because of trouble involving Native Americans. At $2.00 per mile, the 39 men had earned only $8.00 each for nine days of cutting brush, blazing line, and setting a post each mile!

Today, using GPS receivers, the same four miles of line could probably be done by four technicians in a day but at a price that would have amazed the early surveyors and sent Congress into shock. However, most of the public land surveying being done today is retracement of the original surveys, and we must learn to walk in the steps of men like Hutchins and his party.

## THE MANUAL OF SURVEYING INSTRUCTIONS

The rules for the surveys and resurveys of the public lands are set forth in the *Manual of Surveying Instructions*. Editions of the Manual were published in 1855 and reprinted as the Manual of

1871, 1881, 1890, 1894, 1902, 1947, and 1973. When doing any retracement work, it is advisable to try to obtain a copy of the manual that was used when the original survey was made. The survey of public lands is known as **cadastral surveying** and a surveyor that specializes in subdividing the public lands is called a cadastral surveyor.

The Director of the Bureau of Land Management (BLM) determines what are public lands, what lands have been surveyed, and what lands are to be surveyed. He also determines what has been disposed of, what remains to be disposed of, and what is reserved.

The following *General Rules* are from the *Manual of Survey Instructions* and govern the methods we use in retracing the original surveys.

## GENERAL RULES AND THE SYSTEM OF RECTANGULAR SURVEYS

From the foregoing synopsis of congressional legislation it is evident:

First. That the boundaries and subdivisions of the public lands are surveyed under approved instructions by the duly appointed surveyors, the physical evidence of which survey consists of monuments established upon the ground, and the record evidence of which consists of field notes and plats duly approved by the authorities constituted by law, are unchangeable after the passing of title by the United States.

Second. That the original township, section, quarter section, and other monuments, as physically evidenced, must stand as the true corners of the subdivisions which they were intended to represent. They will be given controlling preference over the recorded direction and lengths of lines.

Third. Those quarter-quarter section corners not established in the process of the original survey shall be placed on the line connecting the section and quarter section corners, and mid-way between them, except on the last half mile of the section lines closing on the north and west boundaries of the township, or on other lines between fractional or irregular sections.

Fourth. That the center lines of a regular section are to be straight, running from the quarter section corner on one boundary of the section to the corresponding corner on the opposite section line.

Fifth. That in a fractional section, where no opposite corresponding quarter-section corner has been or can be established, the center line of such section must be run from the proper quarter-section corner. It must be in as nearly a cardinal direction to the meander line, reservation, or other boundary of such fractional section, as due parallelism with section lines will permit.

Sixth. That lost or obliterated corners of the approved surveys must be restored to their original locations whenever this is possible.

**Figure 12-1** Rectangular system of surveys. *Courtesy of the U.S. Department of the Interior, Bureau of Land Management*

## TOWNSHIP GRID

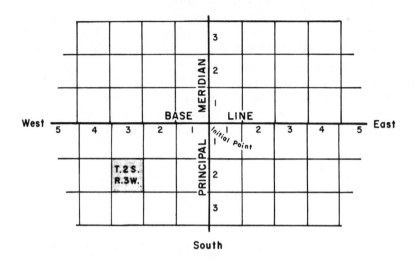

## TOWNSHIP 2 SOUTH, RANGE 3 WEST          SECTION 14

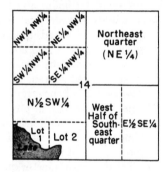

The Manual also states that the basic provisions require that the public lands

> Shall be divided by north and south lines run according to the true meridian, and by others crossing them at right angles, so as to form townships six miles square [and that] ... the townships shall be subdivided into sections, containing as nearly as may be, six hundred and forty acres each.

It further states that

> the excess or deficiency shall be specially noted, and added to or deducted from the western and northern ranges of sections or half-

sections in such townships, according as the error may be in running the lines from east to west, or from south or north.

The system of rectangular surveys fits the basic requirements of the curved surface of the globe (see Figure 12-1).

**TOWNSHIP BOUNDARIES**

In this rectangular plan, the township boundaries are intended to be due north and south or due east and west. The boundaries running north and south are termed **range lines**. The boundaries running east and west are called **township lines**.

The range lines are great circles of earth that, if extended, would intersect at the north pole. This convergency becomes apparent in the measurement of township lines. The convergency is taken up at intervals by the running of standard parallels, on which the measurements are again made full. On the standard parallels (first named "correction lines") there are offsets in the range lines and two sets of corners, standard corners for the lines to the north, and closing corners for the lines to the south. The usual interval between the standard parallels is twenty-four miles, but there were many exceptions in the older surveys (see Figure 12-2).

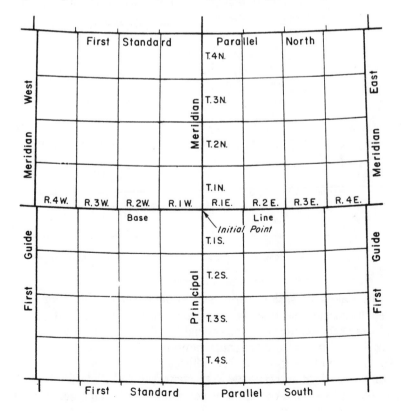

**Figure 12-2** Standard parallels and township boundaries. *Courtesy of the U.S. Department of the Interior, Bureau of Land Management*

**Figure 12-3** Examples of subdivision by protraction. *Courtesy of the U.S. Department of the Interior, Bureau of Land Management*

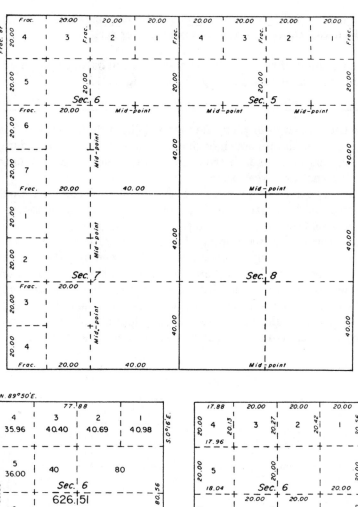

Showing areas.

Showing calculated distances.

**ALIQUOT PARTS**

To make the sections represent "square miles" as nearly as may be, the meridianal lines are run from south to north and parallel to the east boundary of the township. They run for a distance of six + or − miles from the south boundary. These are run and monumented as true lines. The remainder of the section lines are all run by random and true between the established section corners. This produces the rectangular sections, 25 of which contain 640 acres each, within allowable limit. The sections along the north and west boundaries are subdivided on a plan for certain lottings to absorb the convergency and the excess or deficiency in the measurements. These sections provide a maximum number of **aliquot parts** (160, 80, and 40 acre units) or regular **subdivisions** of a section. The remainder are shown as lots whose contents are computed according to the field measurements (see Figure 12-3).

The Bureau of Land Management's service center is located in Denver, Colorado. State offices are located in Alaska, Arizona, California, Colorado, Idaho, Montana, Nevada, New Mexico, Oregon, Utah, and Wyoming. There is also an Eastern States office in Silver Spring, Maryland.

Thirty states have been created out of the Public Lands. Records of the surveys are filed either in the state or with the nearest BLM State office. A list of the states and the locations of records for each state can be found in the *Manual of Surveying Instructions.*

---

If you plan to retrace a public land survey, the first thing you should do is request a copy of the original field notes and the official plat map of the section to be retraced. A thorough search for existing corners should then be made using the calls on the field notes. There are three types of corners.

## TO RETRACE A PUBLIC LAND SURVEY

### An Existent Corner

An **existent** corner is one whose position can be determined either by finding the original monument or by finding the **corner accessories** such as **bearing trees** or objects. If all traces of the monument or its accessories are lost, the description of the point in the field notes, such as line of trees or unmistakable terrain calls, can be used. If none of the original calls can be found, it becomes necessary to search for supplemental survey records that form an unbroken chain of title back to the original monument. If nothing else works, the sworn testimony of a dependable witness that does not have a vested interest in the corner may be used. Have the witness show you where the point is believed to be and take a picture of the witness pointing to the location. Have the person write in the field book how they know where the original monument was located and sign the field book.

### An Obliterated Corner

An **obliterated** corner is one that has lost all trace of the original monument or its accessories, but whose position can be located, beyond a reasonable doubt, by information from interested land

**Figure 12-4** Corner monuments of the public land surveys: A) modern iron post with brass cap, and mound of stone; B) wooden posts, the one on left not much more than a twig; C) wooden post, showing decay at ground line; D) corner monument obliterated, remnants of stone mounds identify corner position; E) corner monument obliterated, evidence of old pits fixes corner position. *Courtesy of the U.S. Department of the Interior, Bureau of Land Management*

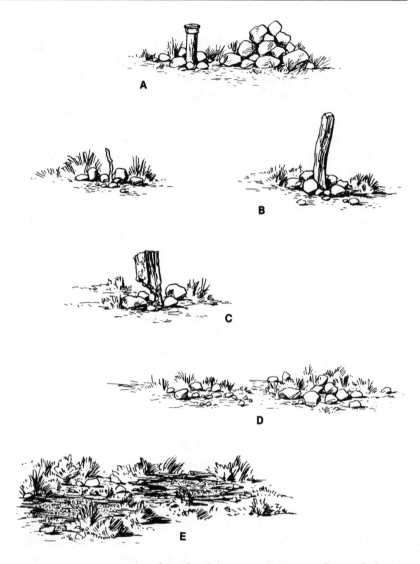

owners, surveyors, local authorities, or witnesses. Recorded evidence of the perpetuation of the corner may be found at the city, county, or state road departments or at the office of the county recorder.

Railroad companies kept excellent field notes of the location of monuments used in surveying the right-of-way lines. The Bureau of Reclamation has exhaustive records of reclamation projects that may have destroyed original monuments after recording their position.

When you have determined the approximate location of an obliterated corner, a thorough search should be made. Mark the supposed location with a lath and begin a search in circles of increasing radii around the lath. Look for anything that does not seem natural to the area. If you find a likely spot, use a shovel to scrape, not dig, the soil away in layers. It is not good practice to destroy evidence by excavating a large amount of dirt in a short time. Look for signs of discoloration in the cleared area. It may be an old rotted post or rust from an old pipe, pin, or gun barrel (see Figure 12-4).

A good search for an obliterated corner is similar to an archaeologist's search for artifacts. If you find evidence of an old post, try to determine the types of wood used and compare the wood to the original notes. If a cedar post was set and you find oak remains, it is probably not the original post, but it may be one set later to mark the position. So, record the position and keep scraping.

If you are sure that you have found the remains of the original monument, tie out its position, place the remains back in the hole exactly where they were found, take a picture of the point showing enough of the surrounding features to positively identify the location of the point and backfill the hole. Set a new corner, using the ties, and mark new bearing trees to the point. A Record of Corner form must be filed for any corner whose position has been perpetuated.

### A Lost Corner

A lost corner is the worst case. A lost corner is a corner whose position cannot be determined beyond a reasonable doubt by any of the methods used for obliterated corners. Do not call a monument lost until everything has been done that can be done to locate it. A lost corner can be restored to its original position only by using one or more found corners and retracing the original surveys between them and the lost corner. Because an existing corner cannot be disturbed, any differences between your measurements and the calls of the original field notes must be proportioned between the found corners.

Proportionate measurement is basically changing your tape or measurements to match those of the original surveyor. Remember that, once accepted, the original measurements cannot be wrong. If your tape says that the recorded distance of 40 chains is actually 39.7 chains, then you must prorate your tape to show the distance to be 2640.00 feet instead of 2620.20 feet. To do this, divide the measurement you have (2620.20) by the record measurement of 40 chains (2640.00) and you find that each foot on your tape is now .9925 feet. Use the prorated measurement to set the intermediate point. If the record distance to the next corner was 41.2 chains (2719.20) you would measure 2698.80 feet with your tape.

**PROPORTIONATE MEASUREMENT**

**Figure 12-5** Double proportionate measurement. *Courtesy of the U.S. Department of the Interior, Bureau of Land Management*

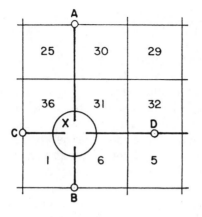

Lost township corner in vicinity of X

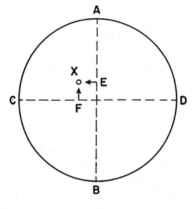

A, B, C, D,—Control corners
E—Proportionate point for X in latitude between A and B
F—Proportionate point for X in departure between C and D

Correct position of X is at intersection of lines extended East or West from E, North or South from F.

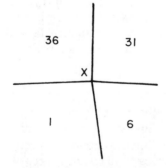

Restored corner showing true direction of township lines

Single proportionate measurement is used along a direct line between two found corners such as a quarter-section corner on the line between two section corners. Any corner on a standard parallel is set by single proportional measurement. All standard corners between township corners are single proportion measured.

A corner common to four townships (lost township corner) or a corner common to four sections inside the boundaries of a town-

CHAPTER 12 / **217**

ship (lost interior section corners) are restored by using **double proportionate measurement**. With double proportionate measurement, the distances between four found corners, two in each cardinal direction from the lost corner, are used to find the distance to the point (see Figure 12-5).

The single proportionate measurement between the two found corners north and south of the missing corner is used to set the north/south point on line between the found corners. A temporary point is set on the line at the proportionate distance. The distance between the nearest found corners to the east and west is measured. A temporary point is set on line between those corners at the proportionate east/west distance. A line is set in the cardinal direction (N, S, E, W) from each of the temporary stakes. The new corner is set at the point of intersection of the two lines (see Figure 12-5). There is a definite order of preference for standard corners. Parallels will be given preference over townships, townships over sections, sections over quarter corners. The interior quarter corner (center of section) will be set at the intersection of the lines connecting the four quarter corners.

THE LATITUDINAL
CURVE OF A LINE

On east/west township lines and on standard parallels, the proper correction must also be made for the latitudinal curve of the line. There are two commonly used methods to establish the distance to be set off from the line to put the points on the curve. The **tangent** method is useful in open country where brushing is not necessary (see Figure 12-6).

In timbered country, the **secant** method is probably better as it does not need the long offset distances required by the tangent method (see Figure 12-7).

The standard field tables give the offsets for any degree of latitude from 25 to 70° north, and for each half mile from 0 to 6 miles. The standard field tables also give the convergency of the meridians along the 6-mile distance from township corner to township corner.

When the original surveys were made, all **navigable** rivers as well as nonnavigable rivers more than 3-chains wide, were given a **meandered line** on both banks at the ordinary mean high water mark. All lakes more than 50 acres in area were also meandered. A **meander** corner was set whenever a standard, township, or section line intersected the bank of a meanderable body of water. Islands above high water mark were also meandered. Lost meander corners on a meanderable body of water that was marked on both banks will be restored by single proportionate measurement between the nearest standard corners found on each side of the lost meander corner.

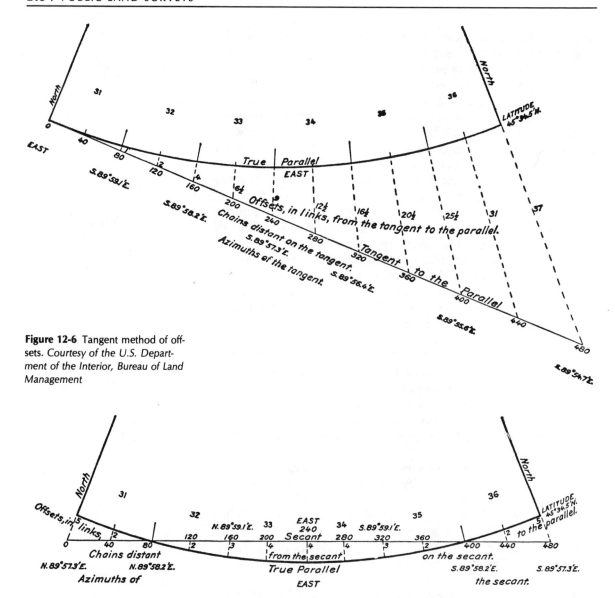

**Figure 12-6** Tangent method of offsets. *Courtesy of the U.S. Department of the Interior, Bureau of Land Management*

**Figure 12-7** Secant method of offsets. *Courtesy of the U.S. Department of the Interior, Bureau of Land Management*

Because of the convergency of meridians, a closing corner was set at the true point of intersection of the guide meridian with the standard parallel. Closing corners were also set on the north of townships ending on a base line or parallel. At the intersection of the survey lines with Native American reservations, land grants, and state boundaries, a closing corner was also set. A closing corner provides control for direction only and is intended to end at the true intersection with the line closed upon. A lost closing corner will be restored on the true line between the nearest standard corners to each side using single proportionate measurement and set over to the true line. In some cases a line is terminated with measurement in one direction only. In this case, the lost corner will be restored by record bearing and distance from the nearest regular corner. For a measurement in one direction only, it is advisable to retrace another line run by the same surveyor and correct your chain to his by determining an index error for alignment and distance.

Once the original corners have been found or reestablished, the subdivision of the section can begin. The quarter-quarter section corners are set on the line between the section and quarter-section corners at the halfway point. An exception is in sections closing on township boundaries, where they should be set at 20-chains proportionate measurement from the quarter corner. The center of the quarter section is established by running straight lines from the opposite quarter-quarter corners and setting the point at the intersection of the lines run. The remaining aliquot portions are subdivided in the same way. The smallest aliquot lot is usually a quarter-quarter-quarter or $\frac{1}{64}$ of a section containing 10 acres more-or-less and measuring 660 ft. to a side.

The original surveys of the public lands are of interest to the technician because the methods used by the original surveyors may give insight into the discrepancies between the record and measured distances found in retracement surveys.

**THE SUBDIVISION OF A SECTION**

Each survey has a point of beginning; for rectangular surveys, that point is the principal meridian and base line (see Figure 12-8). Learn the name of the controlling base and meridian for your area. From that point a principal meridian is run north and south along the true meridian. Regular quarter sections and section corners are set at 40-chain (half mile) intervals.

Meander corners are set at the intersection of meanderable bodies of water. The line was usually run using a solar compass or a solar transit in later years. Two independent measurements of the line were made but only the mean distance was recorded. The allowable closing error was 7 links in 80 chains or 4.62 ft. in a mile. The maximum angular error was 3 min. from true north. Another line was run east and west with regular corners being set at 40 and 80 chains.

The offset to the true latitudinal curve was made by solar, tangent, or secant method. The accuracy required was the same as

**THE PRINCIPAL MERIDIAN AND BASE LINE**

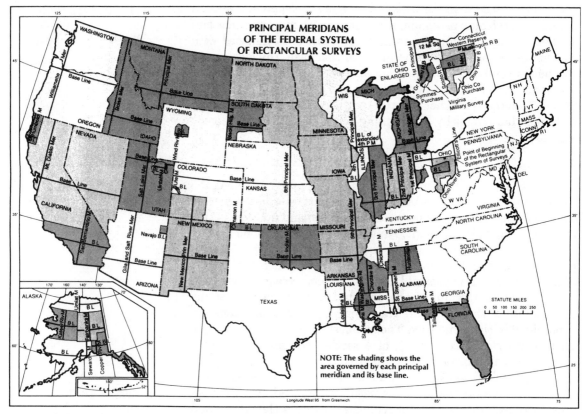

**Figure 12-8** Principal meridians. *Courtesy of the U.S. Department of the Interior, Bureau of Land Management*

for the meridian. A standard parallel was run east and west every 24-mile segment along the principal meridian. A closing corner was set at the true point of intersection with the next standard parallel north. The distance to the standard corners east and west was measured and recorded in the field notes. Each 24-by-24-mile block is called a "quadrangle." Whenever possible, townships of 6-miles square were surveyed from the south successively to the north. The meridianal boundary lines have preference and were run from south to north on true lines. Quarter and section corners were established, and a closing corner was set on the parallel. Any excess or deficiency was placed in the last half mile. The latitudinal township line was run on a random line, from the old to the new township line, and corrected back to true. Any excess or deficiency was put in the last half mile. It becomes evident, from this, that any error in a township will be put in the north and west sections in the last 20 chains. Whenever it is necessary to do a retracement survey in the north half mile of sections 1, 2, 3, 4, 5, and 6, be careful to watch for excess or deficiency in the closing distance. This is also true for the west half mile in sections 6, 7, 18, 19, 30, and 31.

The practice of marking the line of survey with "hacks" or "blazes" (see Figure 12-9) was modified by the National Environmental Policy Act of 1969. The Act allows discretion in the esthetic value of marking trees, especially on private land. Line trees, if marked, will have two hacks, or notches, cut on each side facing the line. Trees within 50 links of the line are blazed on two sides, quartering toward the line. These marks last as long as the tree stands. The surface of the tree will grow over the cuts but traces will show on the bark for many years. Do not cut into a standing, living tree to look for the marks unless it is absolutely necessary. Disease and insects may get into the wound and cause the tree to die, destroying the evidence of the marked line.

Another way to retrace line is from the calls on the field notes. The following features were recorded in the field by the survey party:

1. The bearing and distance of each line run.
2. The species and diameter of each bearing tree, all bearing objects, and witness corners and their bearing and distance from all corners.
3. The size, material, and markings of the corner set as well as its accessories.

**Figure 12-9** Tree markings: the marks found on a line tree may be a blaze, a hack, or both. *Courtesy of the U.S. Department of the Interior, Bureau of Land Management*

4. The species and size of each line tree as well as the distance from the corner line.
5. The type of objects intersected by the line such as boundary lines or reservations, town sites, and private claims. The distance to and bearing of the boundary line and its nearest corner, the center line of railroads, canals, ditches, power lines, or other rights of way must be recorded. Also, record changes of terrain and vegetation, ascents and descents of slopes and the direction of slope, plus landmarks such as cliffs or ridges.
6. The size, direction of flow, and distance to the center line of smaller streams or both banks of larger streams and the course of the bed of the stream.
7. The type of surface traversed such as level, rolling, broken, hilly or mountainous.
8. The soil type, such as rocky, sandy, loam, etc.
9. The type of timber and undergrowth.
10. The types of land such as upland, swamp, or overflowed land.
11. The location of streams, springs, or water holes, plus the kind of water and the direction of flow.
12. The location of lakes and ponds and a description of the banks and streams flowing into and out of them.
13. Man-made improvements such as fields, houses, mines, or other man-made objects such as U.S. location monuments.
14. Mineral outcroppings, beds, or bodies.
15. Roads and trails and the direction of travel.
16. Waterfalls, rapids, and their height of fall.
17. Rock quarries, ledges, and type of rock.
18. Natural formations of interest, such as fossils, petrified logs, or archeological remains such as dwellings, forts, or mounds.
19. Magnetic declination and areas of local attraction.
20. A general description of the land crossed and its fitness for settlement.

These things were important to the prospective settlers and can be useful in a retracement of the public land surveys.

**THE TYPE OF MONUMENT**    The type of monument set is described in the field notes and must be of approved material and size. The *Manual of Surveying Instructions* that was in effect at the time of the original survey gives the general requirements for the various types of monuments. The description will give the type and size of the monument, the depth set into the ground, and the markings on the monument.

The standard monument in the 1973 *Manual of Survey Instructions* is the regulation post made from zinc-coated iron-alloy pipe with a 2.5-in. outside diameter and 30-in. length. The bottom end is split and spread to form flanges. A brass cap is securely fastened to the top. A brass tablet 3.25 in. in diameter with a 3.5-in. stem is used for locations in solid rock. If the point falls on rock that is too hard to drill, a chiseled cross is cut at the exact point and a stone monument is built above the "X."

**CORNER ACCESSORIES**

If the corner point should fall in the same position as a tree, the tree will be marked and the accessories set to reference the position of the tree. Corners that fall in cultivated fields are probably buried and a guard post set. Bearing objects or trees within a "reasonable" distance are used and "accurate bearings to one or more available distant objects are recorded."

In locations where it is impossible to occupy the actual corner point, a **witness corner** is set on the line. It must be as close to the corner as possible, at least within 10 chains. If the corner falls in an unimproved roadway, a marker is buried at the corner point and at least two reference monuments are set outside the roadway. On improved roadways, a point is set on the surface and two to four reference monuments are set.

A **witness point** is set "to perpetuate an important location more or less remote from, and without special relation to, any regular corner." The marks on the corners must be clear and easily read. Steel stamps are used on brass caps, whereas a timber scribe is used on trees. Notches and grooves are used on stone monuments to indicate the distance in miles to the east and south boundaries. The commonly used marks are

AM – Amended Monument
AMC – Amended Meander Corner
AP – Angle Point
BO – Bearing Object
BT – Bearing Tree
C – Center
CC – Closing Corner
E – East
EC – Electronic Control
LM – Location Monument
M – Mile
MC – Meander Corner
N – North
NE – Northeast
NW – Northwest

R – Range
RM – Reference Monument
S – Section
SC – Standard Corner
SE – Southeast
SMC – Standard Meander Corner
SW – Southwest
T – Township
TR – Tract
W – West
WC – Witness Corner
WP – Witness Point
¼ – Quarter Section
¹⁄₁₆ – Sixteenth Section

For a complete listing of the proper method of marking the corners see "Marks on Corner Monuments" in the 1973 *Manual of Surveying Instructions.*

The "corner accessories" consist of:

1. Bearing trees or other natural objects such as "notable cliffs and boulders, permanent improvements, reference monuments."
2. Stone mounds.
3. Pits and memorials in the ground.

"Bearing trees" are trees that are marked on the side facing the corner with the letters BT. They also show section, township, and range scribed in a vertical line, reading downward. The blaze is approximately one axe-handle long and one axe-blade wide, ending one axe-blade above the root swell of the tree. Metal plates have been used, when possible, since 1969. One tree in each section should be chosen, if possible, and the appropriate section marked on the blaze. A chiseled cross is cut into rock cliffs or boulders along with the letters BO to mark a "bearing object." A bearing and distance call to "the southwest corner of the foundation of Smith's house" may be used, but no marks are set on private property without permission of the owner. If there are no trees or bearing objects and where mounds or pits would not be practical, a "memorial" of glass, a marked stone, charred wood, or charcoal may be deposited alongside the monument. When replacing an old monument with a new one, the old monument may be deposited alongside the new monument.

A "stone mound" may be used if there is native stone available in the area. The mound should be at least five stones and no smaller than 2-ft. wide by 1.5-ft. high. The nearest point of the stone base should not be closer than 6 in. to the monument. "Pits" 18-in. wide and 12-in. deep, with the nearest side 3-ft. away from the corner, may be used in open country when nothing else is available.

---

The use of **photogrammetric maps** to set the control for both original corners and to locate corners on retracement surveys is becoming more common. With the use of helicopters to access remote sites and the ability to measure between points that are not intervisable with GPS, the cost of surveying large areas of the public lands is much less than the cost of conventional surveys. Using existing high order geodetic horizontal and vertical control network monuments and state plane coordinates for positional control of the aerial premarks, the corners can be set by protraction. For retracement surveys, existing corners can be plotted on an aerial **orthophoto** of the search area and the direction and distance to missing corners scaled in. Aerial premark panels can then be set fairly close to the theoretical position of the missing corners.

**Analytical phototriangulation** is then used to compute the state plane coordinate values for the control points. By inversing between the control point and the theoretical position coordinates of the missing corner, a bearing and distance can be computed. The field crew can then make a search of the most likely position of the corner and make a determination about the condition of the corner: found, obliterated, or lost.

If the corner is determined to be lost, it can be reset from the control point using approved methods. Because a traverse has not been run through the corner, the accuracy must be given as a radius of error or as a positional tolerance.

The use of astronomic azimuth will become a necessary skill for the surveyor to master if another control point is not visible from the previous control point. Using modern survey instruments and methods, a proficient survey technician should be able to achieve an azimuth error of not more than ±5″. An error of 5″ will give a positional error of .03 ft. in a quarter mile. A correction from astronomic north to geodetic north may be necessary for long sights because of the variation of gravity. Called **local anomalies**, these variations cause a deflection from the vertical. This can be compensated for by use of the "Laplace correction" that is given for NGS stations throughout the country.

It is evident from the differences between the number of men needed by Hutchins to do his surveys and the number of technicians needed to set corners today that the level of training needed to do the work is much higher. Only by continued education and a

**THE USE OF PHOTOGRAMMETRY**

great deal of practice will the technician gain the skills necessary to become an expert cadastral surveyor. A good library of books for reference on the subject will be a big help as will attending seminars put on by your state and land surveyors' associations.

## Review Questions

1. In what directions are range lines run?
2. In what direction are township or tier lines run?
3. How do you correct for convergency in townships?
4. Running north from the south boundary of the township, in which township is the shortage or excess in distances placed?
5. On which boundary of the township would you expect to find corrections or closing corners?
6. How far apart should standard parallels be run?
7. How many acres are in a standard section?
8. Where would you expect to find government lots?
9. Make a drawing showing the aliquot parts of a standard section.
10. Define an existent standard corner.
11. Define an obliterated standard corner.
12. Define a lost standard corner.
13. Where should single proportionate measurements be used?
14. Where should you use double proportionate measurements?
15. On east/west township lines and on standard parallels, a correction must be made for the latitudinal curve. Name the two methods used and give the advantage or disadvantage of each.
16. Where are meander corners set?
17. How is the center of section corner set?
18. What are the names of the controlling bases and meridians for your state?
19. What are corner accessories?
20. How are bearing trees marked?

# Legal Descriptions

## Objectives

After completing this chapter, the student should be able to:

1. Gather evidence to be used by a licensed surveyor for boundary location.
2. Define evidence:
   A. Written,
   B. Real,
   C. Parole.
3. Recognize a monument:
   A. Artificial,
   B. Man-made.
4. Describe the proper method to search for a monument in the field.
5. Describe how to preserve any evidence found when searching for a monument in the field.
6. List, in descending order of importance, the calls for boundary monuments shown on the deed.
7. Differentiate between "metes and bounds" and "subdivision" descriptions and give examples of each.
8. State the general requirements of a deed.
9. Define: Acquiescence, adverse possession, act of nature, eminent domain, and prescriptive right.
10. Describe how to research a deed in your particular city and county area.
11. Relate the steps to be followed in surveying a lot in the field from arrival at the job site to setting of the corners.

12. Define: Riparian, littoral, navigable and nonnavigable bodies of water, accretion, reliction, avulsion, thread of a stream, and thalweg.
13. Identify the variables in an "along the road" call in a deed.
14. State the problems that can be caused by an "of" description in a deed.
15. State the problems that can be caused by an "area conveyance" in a deed.
16. Distinguish between "junior" and "senior" rights in a land subdivision dispute.
17. Define "ambiguity."

## LAND GRANTS

In chapter 12, you learned that an "aliquot" description of a piece of land makes an excellent legal description. However, not all land can be described in relation to its position in a section. Twenty states were not a part of the public lands that were divided into townships and sections. Several other states have lands that have titles going back to the land grants made by the nations that originally had possession of the land from the Native Americans.

The lands still possessed by the Native Americans are not part of the public lands. The systems used to describe the boundaries of these excluded lands vary from local rectangular systems to **metes and bounds descriptions.** There were more than twenty different systems used in Ohio alone. The French made land grants along the Mississippi River and its tributaries in the fashion used in France. This system granted a certain number of **arpents** of land (a French measurement) along the river and extending back from the water 40 to 80 arpents in some vague direction.

The Spanish held vast territories in the west and granted titles to the settlers who were honored by the Mexicans when they gained independence from Spain in 1821. When the United States gained possession of the lands, the Spanish and Mexican grants were honored. The Treaty of Guadalupe-Hidalgo that recognized existing land grants would continue under the Mexican law, which was in force at the time of the grant. Laws governing minerals and water were different under Mexican law, where title was retained by the crown.

The unit of measure was the **vara,** which differs from place to place. In Texas, the vara is 33.3 inches. When Texas joined the Union in 1845, it kept its own land system.

## EVIDENCE

At the time of Mexican independence, only 20 privately owned ranchos existed in California. By 1847, the number had grown to 800, ranging from 28 to 226,000 acres. Because of the many different systems and the general lack of sufficient descriptions of the boundaries of ownership, it falls to the courts to decide the legal boundaries of the land in cases of disputed ownership. Your job as a survey technician is to gather **evidence** to be used by the licensed surveyor if the case is presented to the courts for decision. That evidence can be found in titles, on maps, in public records, and by testimony of owners. Evidence is also found on the ground as shown by indications of possession and existing monuments. *Black's Law Dictionary* defines evidence as, "Testimony, writings, material objects, or other things presented to the senses that are offered to prove the existence or nonexistence of a fact."

In the field, the technician is the eyes and ears of the land surveyor (LS) and must know what to look for. The information on what you find in the field must be transmitted clearly to the office. The best way to transmit this information is by clear and complete field notes. In this chapter, you will learn what to look for when in the field and why it is important as evidence.

**Written evidence** is found in titles, maps; public records, and documents that pertain to the location of the boundaries of the property being surveyed and any encumberances of record on the property.

**Real evidence** is what you find on the property and around it such as monuments, fences, roads, signs of continued use, marked trees, buildings, underground utilities, and anything else you feel may be of use in determining the boundaries of the property. It is always better to gather more evidence than is needed than to miss some important feature that to your embarrassment may come up later.

**Parole evidence** is given by witnesses who actually saw the location of real evidence when it existed or who can give an explanation of why there is a conflict in the description. A witness may have been present at the time the original survey was made and can give testimony regarding where the points were or what the original parties to the deed said or did.

A "monument" may be a natural object such as a tree or rock. Streams or creeks may mark the boundaries of the property as it is now. An "artificial monument" may be a stake, a pipe or gun barrel, a marked post, a concrete marker, a rebar with or without a plastic cap, or even a road. An original monument called for in the deed or other writing is the most important evidence of the intent of the original parties. When there is a call to a monument on the legal description, it is very important that it be located on the ground. A thorough search should always be made for any called-for monument before it is presumed to be lost. If no sign of the

**MONUMENTS**

monument exists on the surface, a careful excavation should be made when looking for any trace of the monument or any other object that may have been set in its place. Shave the dirt away in thin layers and look for any signs of something different from the surrounding soil. Traces of wood or rusted metal or even the discoloration of the soil may indicate the former location of a monument (See discussion in Chapter 12 and Figure 12-4).

If you find something, preserve the evidence by tying out the location and, if possible, take a picture of the evidence in place. If you decide to dig deeper, save the evidence of the point and try to replace it in the same position when refilling the hole. If it is decided that the location of the monument has indeed been found, use the ties to set a new monument in its place. Then make a record of what was found and what was set in its place.

The chain of history of a monument is very important in establishing the location of the point. If more than one monument is found in the vicinity of the point, an effort must be made to decide which is the best evidence of the true location of the original survey. Normally, natural monuments take precedence over artificial monuments and artificial monuments take precedence over written monuments.

An uncalled-for monument may take precedence by "common report" if a number of surveyors have used the uncalled-for monument over a period of years. When it is difficult to tell which monument is best, the one that comes closest to fitting line and distance or area calls should be used. Be sure to make note of any other monuments found and their bearing and distance from the one used. If the monuments have the registration number of a LS or professional engineer (PE) stamped or attached to them, be sure to record the number in the field notes.

After original deed monuments, in order of importance are monuments called for on maps or other documents that pertain to the property being surveyed. These are the called monuments on adjoining property or boundary lines that are marked on the ground by some physical evidence. Next are references to a map or deed boundary that may be located. Next come bearing or distance ties to record monuments not in close proximity to the property being surveyed, followed by the distances on the boundary without relation to monuments. Next in order are the bearings or angles between lines (except in Texas). Last are calls for area, such as "five acres more or less." This includes more specific calls such as "the west 50 ft. of lot 10" or "the south 15 acres."

## TYPES OF DESCRIPTIONS

There are two types of descriptions: the "metes and bounds description" and the "subdivision description." In the metes and bounds description, a direction of travel around the parcel is described in terms of "metes" or measurements of distance and bearings. "Bounds" are restrictions in the direction of travel by encountering other boundaries or monuments. "Beginning at the southwest corner of Johnson's Ranch thence N 00-15W, a distance of 200 feet to a point, thence S 88-25 W, a distance of 350 feet to a 2 inch iron pipe," is a metes and bounds description.

In a subdivision description a call to a map is used. "Lot 5, Block 3, of Jones' Addition to Whatcom, as recorded in book 9, page 36 . . . " is one type of subdivision description. A call to an aliquot portion of a section, "the west ½ of the SW ¼, of the NE ¼, section 6, . . . " is another subdivision description.

When a distance and bearing are given, along a course of a metes and bounds description, together with a call to a monument, "thence S 88-25 W, a distance of 350 feet to a 2 inch iron pipe . . . ," the monument will hold precedence over the distance and bearing. On calls for a natural or man-made monument, "the east line of Highway 90" or "to the Merced River," the monument as it now exists will be the end of the line, not the distance call.

---

**TRANSFERRING OWNERSHIP**

A **deed** is the most common method of transferring ownership or interest in real property from one party to another. The general requirements of a deed are

1. Competent parties
2. Proper subject matter
3. A valid consideration
4. Written form
5. Sufficient wording
6. Reading before execution
7. Execution, signing, sealing, and attestation
8. Delivery

Property can also be transferred through **unwritten conveyances.** There are two types of transfers of real property that do not require written conveyances. One is the transfer of property by actual oral agreement or by implied agreement by the conduct of the parties. The other is the involuntary transfer of property by the transferors (seller) arising out of adverse use by the transferees (buyer) without disturbance by the owners over an extended period of time. Land may be lost by **acquiescence,** by **adverse possession,** by **act of nature,** by **eminent domain,** and by prescriptive right without written transfer. These are the reasons the technician must be alert when on the job for signs of use by parties other than the client.

Acquiescence infers an agreement between two parties although none may have in fact existed. If two adjacent landowners, over a long period of time, accept, as a boundary marker, a hedge, a fence, or some other physical monument, by their actions the monument may become the true boundary. The statute of limitations for boundary disputes varies from state to state so be sure to know the law in your state.

Adverse possession is the legal term for what is called "squatters' rights" by some people. This was originally intended to prevent the absentee land owner from appearing after a long period of time and forcing the people who had developed the land to leave the homes and farms that they thought were theirs. There are ten elements that must be satisfied, in most states, for adverse possession to ripen into fee title.

1. Actual possession: cultivation, fencing, farming, improving, or otherwise showing signs of possession.
2. Open and notorious possession: the true owner must be able to observe the possession. It must not be secretive or hidden.
3. Claim of title: the claim of title may be defective or mere color of title, but some written right to the land must be present. A squatter cannot claim adverse possession.
4. Continuous possession: the person claiming adverse possession must have been continually on the land for the statute period of time. It is presumed by law that possession resides in the true owner. If someone else is not in actual possession, the land is in possession of the true owner. If the claimant leaves the land, the claim ends and must be started over to ripen into title. Short periods with an intention to return do not cause an interruption in possession.
5. Hostile or adverse possession: the possession must act against the interests of the true owner. Hostile means to the exclusion of all others but does not imply ill will against anyone. If the true owner gives permission to use the land the adversity ends.
6. Exclusive possession: the possession must not be shared with the true owner. Exclusion can be shown by threat of force or by legal action, by declarations to hold the land exclusively, or by refusing to let the owner or his/her agents enter on the land.
7. Possession for the statute period of time: most states have a statute period of from seven to twenty years.
8. Color of title: any written instrument that is not invalid on its face may show color of title if it purports to convey title to the land in dispute.
9. Taxes must be paid: payment of taxes is evidence of a claim in some states but not required in all states.

10. Good faith: some states require that the claimant believe that a good title to the land is in their possession and they have acted accordingly to take possession of the land.

Land may be gained or lost by an act of nature: either **accretion**, the slow and imperceptible adding to the land by deposit of water-borne material or by **erosion**, the slow and imperceptible wearing away of the land by water action. Title to the land moves with the bank of the stream in its natural wanderings. One person's loss is another's gain and both stand an equal chance of gaining or losing land by acts of nature.

Eminent domain is the right of the sovereign (government) to take private property for public use. The Fifth Amendment to the Constitution states that no person shall be deprived of life, liberty, or property without due process of law, nor shall private property be taken without just compensation. The right of eminent domain has been given to cities, counties, public utilities, highway departments, parks, school districts, railroads, pipeline companies, and other public and private corporations that serve the public need. Most road and power easements are obtained by condemnation of private property. When doing a survey of a piece of land, always be sure to check for easements over the land that may not have been used yet. These are called reserve easements and are usually for roads or public utilities.

Prescriptive right is the right to an easement across another person's property. It is similar to adverse possession but does not carry title to the land, only an easement to use the land in question. A child cutting across an empty lot to go to school is one example. If you have used the trail for the statute period of time and have not been barred from its use by the owner, a right of use may be sought by legal action. Permission to pass is revocable. Signs are used by railroads and many public buildings to prevent the continued use of their property from ripening into prescriptive right. A gate across a road, that is occasionally locked to bar passage, is also an interruption of the use of the road by other than the owner. Be sure to look for signs of continued use by others across any property that you are surveying and report it to your employer.

Exactly how to locate a written title boundary on the ground is a difficult thing to learn. As every description is different, so each survey is different and each will have its own particular solution. The boundaries are located by the best possible evidence available to the surveyor. The amount of time available to do the research is not infinite so each person will probably find slightly different

## LOCATING A WRITTEN TITLE BOUNDARY ON THE GROUND

record information. Usually, the research will begin with a deed or title insurance policy issued to the owner of record. The names of all adjoiners called for in the legal description can be found in deed or title policy. Copies of the adjoining conveyances and related maps can be obtained from the local courthouse or office of the county recorder. Copies of records of survey can also be obtained. The office of the county surveyor or city engineer may also be a source of much information on the adjoining properties. As not all deeds or surveys are recorded, it may take a little detective work in the area to uncover unrecorded conveyances and surveys. Copies of the surveys of the public land corners can be obtained from the sources listed in chapter 12. Be sure to check for court rulings that may have changed the boundaries of the land. There are as many as sixty-six different places to get property line information. Because you cannot cover all the possibilities each time, just do the best you can.

Once the decision has been made on which point or points to start from, the field party can begin retracing the original survey, if one was made. As you learned earlier in this chapter, the recovery of original monuments called for in the survey and legal description is the most important. An undisturbed, original monument that is called for determines the exact position intended by the original parties. Any restored monuments that are established by measurement alone have a positional error. If two original monuments are found, it is possible to determine an index error between the original measurements and your own measurements. Use the index distances when looking for other points set by the same surveyor.

If the survey is part of the subdivision instead of a metes and bounds survey, there may have been a simultaneous conveyance of all the lots at the time the subdivision was recorded. In the case of a simultaneous conveyance, each lot will receive a proportional part of the entire subdivisional block.

First, the boundaries of the subdivision must be established by adjoiners or by a metes and bounds retracement. If the streets have been improved, monuments at the center line intersections are usually set from the original property corners. The center line monuments can be used as reference monuments to the lost property corners if shown on the plat. The distances and angles between the four adjoining street center lines should be measured. If there is a significant difference between the recorded and measured distance of the bearing, a check should be made to see if the monuments may have been set at an offset distance to allow for utilities in the street. Sometimes a sewer manhole may have been installed at the center line intersection and punch marks made in the rim to perpetuate the location of the center line intersection. If no monument is visible on the surface, the road may have been repaved.

Check the center line ties in the curb or sidewalk and contact the city or county surveyor for recorded ties. There may be concrete or PK nails set close to the location as crossties. Use your plumb bob string to line between the opposite nails and make a mark on the paving at the intersection point.

If you have a good metal detector, it may be used to locate a buried point if the point contains enough metal to detect. If it is necessary to dig a hole in the roadway to find the monument, set crossties at the location of the point, if found. Carefully refill the hole and tamp down the material. Do not leave an open hole in a roadway that might cause an accident.

Some center line monuments are set in wells with a cast-iron cover. Dirt will work down around the cover and cause it to become stuck. Do not try to pry the cover off with your plumb bob; you will just break off the point. Use a hammer to pound on the cover and loosen the dirt. Then use a cold chisel to raise the cover enough to lift it out. A large spoon or a trowel can be used to clean out the dirt and a paper cup to bail out the standing water. A small mirror is useful to shine light into the well to find the point in the monument.

After the center lines are determined, the side lines or property lines are set at record distance from the center line. The half street distances should be measured at right angles to the center line. The width of the street is not prorated with the rest of the block but is set at the full dedicated width. Once the streets are taken out, each of the rest of the lots gets a proportional part of the remainder. If the block is 1-ft. short and there are ten lots, each lot is reduced by one tenth of a foot. Any excess or deficiency is confined to the block where found and not carried over to the next block.

If an original monument is found in the block, the distance cannot be carried across it but each side of the monument receives its prorated share of excess or deficiency. If no distance is given for the end lot on the block, all of the error is placed in that lot instead of being distributed among the other lots. If the end lot, adjacent to the subdivision boundary, calls for a larger distance than is available, it cannot extend beyond the boundary line of the subdivision. If the end lot does not reach the boundary, the excess is not given to the end lot but reverts to the subdivider.

## RIPARIAN AND LITTORAL OWNERSHIP

If one of the boundaries of the land being surveyed is a body of water, there are special methods of determining the limits of ownership. If the land is adjacent to a river or stream, the ownership is called **riparian**. If the land is next to a lake or tidewater, it is called **littoral** ownership.

Water that is capable of supporting commerce is considered to be navigable and the ownership of the bed is retained by the government. The upland owner bordering on navigable streams owns to the high water mark. The high water line is the line of vegetation below which the normal rise of the water prevents the growth of upland bushes and trees.

On nonnavigable streams, the upland owner owns to the center of the stream. The problem arises when you try to define navigable and nonnavigable or the center of the stream. Different states have different definitions of these terms, so be sure to learn the court's interpretation for the state in which you are surveying.

As you learned earlier, a riparian owner can gain or lose land by the action of the water boundary. If land forms along the bank of a river, lake, or stream by the slow deposit of **alluvion** through the process of accretion, the landowner gains the land. If the land is uncovered by the slow recession of water, as in a lake drying up, the process is called **reliction**. If the land is suddenly torn away, as in a flood, by the river changing its bed or forming a new channel, the ownership does not change but remains as it was. This process is called **revulsion** or **avulsion**. If the land is gradually worn away by the action of water, the process is called erosion and the land owner loses the land. The thread of a stream is a line midway between the banks when the water is at its normal stage. The **talweg** or **thalweg** is the center of the deepest or lowest part of the main channel of a river or stream.

## "ALONG THE ROAD"

If a boundary of the land is a highway, the ownership is similar to that of a riparian owner. If the original grantor owned the land to the center of the highway when the original transfer of title was made and the highway is abandoned, title to the land underlying the highway reverts to the title holder.

If the title reads "along the road" and is not limited by a call to the "edge of the road" or "the sideline of the road," title is **conveyed** to the center line of the road if the road is abandoned. Unless the language of the deed specifically calls for the ownership to end at the "side" or "edge" of the road, the courts have held that title was intended to be conveyed to the center of the road. The granting of an easement does not carry fee title to the land underlying the easement and ownership of the land remains with the grantor of the easement.

When land is conveyed by an "of" description, the possibility of an **ambiguity** exists. A description that reads "the westerly 50 feet of lot 26" conveys 50 feet of land; the question is which 50 feet? As seen in Figure 13-1, the 50 feet could be measured from the side of the street or from the center, if title is owned to the center of the street. The courts have held that the side line of the street would

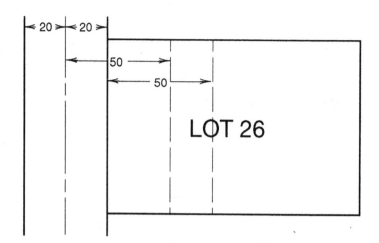

**Figure 13-1** Westerly 50 feet of lot 26

be the proper place from which to measure because of the fact that the grantor has no right of occupancy of the street. Lot lines do not change even if the street has been abandoned. The additional land would be added on to the original lot width if the street were abandoned.

A different meaning is applied with a metes and bounds description. Ownership to the center of the road is implied, unless excluded by deed, and "of" measurements are made from the center of the street if that street is a boundary of the metes and bounds description. The distances in "of" descriptions are made at a right angle to the boundary of the line from which the measurement is being made. In a proportional "of" conveyance, such as the North one half, the measurement is presumed to be based on area unless otherwise stated in the deed (see Figure 13-2).

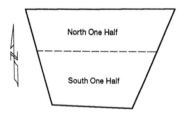

**Figure 13-2** North one half by area

## JUNIOR AND SENIOR RIGHTS

In area conveyances, the ambiguity comes from the land not laid out in a **cardinal direction**. In Figure 13-3, which half is intended to be conveyed? If the description reads "the northwest one half as measured along the southwesterly line" the ambiguity is removed. If a distance or area is added to the description, such as "the westerly 100 feet" or "the westerly 5 acres," a **Junior** or **Senior** right may arise. A person cannot grant title to the same land twice. You cannot sell what you do not own. Suppose title to "the westerly 100 feet" is given to one person and "the easterly 100 feet" to another person at a later date. If the grantor does not have the full 200 feet to grant, the first grantee gets 100 feet and the second gets the remainder (see Figure 13-4).

The first buyer has senior rights to get the land deeded by the grantor. The second buyer gets any remainder, if the land is short of the granted distance. If the land is longer than the granted dis-

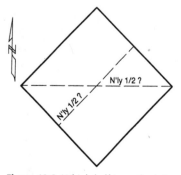

**Figure 13-3** Which half is northerly?

**Figure 13-4** Junior and Senior rights shortage

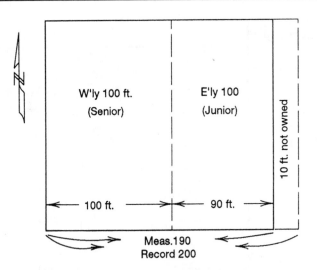

tance, the excess is not given to the second buyer but is retained by the seller (see Figure 13-5A). If the deed reads "except the westerly 100 feet" then the junior owner gets any excess land owned by the grantor (see Figure 13-5B).

Remember, it is the survey technician's job to keep careful and accurate notes on each survey job. If the boundary is disputed in court, the judge has final say and can ignore everything in making his or her decision. To win in court, the surveyor must try to follow these guidelines and then explain the logic to the judge in an attempt to show why this logic is *better* than that used by the surveyor for the other party.

**Figure 13-5** A) Senior and B) Junior rights

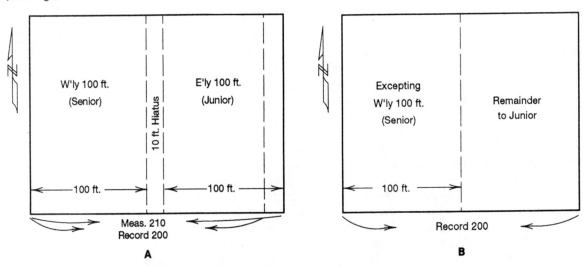

# Review Questions

1. What is evidence?
2. Why are good field notes so important?
3. What is written evidence?
4. Monuments, fences, roads, signs of continued use, and marked trees are what kind of evidence?
5. How much evidence is necessary to prove a boundary?
6. What is parole evidence?
7. What is a monument?
8. Name some types of artificial monuments.
9. What is the most important evidence of intent of the original parties to the deed?
10. Why must you be very careful when searching for remains of a monument?
11. How can you preserve evidence of the location of a monument?
12. What is a metes and bounds description?
13. What are subdivision descriptions?
14. Which has greater significance as evidence – a call for distance, a call for bearing, or a monument called for in the deed?
15. Name three ways a property may be gained or lost by unwritten conveyances?
16. Name five types of adverse conveyance?
17. Give an example of each of the five adverse conveyances in question 16.
18. When measuring between the original monuments, which measurement is the correct distance: the original surveyor's recorded distance or the distance you measure?
19. Should the street widths be taken as shown when prorating the measurements in a block?
20. What is the ownership of land next to a river or stream called?
21. What is the ownership of land next to a tide water or lake?
22. Define high water mark along a navigable stream or river.
23. What do you call the process of slow wearing away of the land by water?
24. What do you call the sudden tearing away of the land by water?
25. What is the center of the deepest or lowest part of the main channel of a river or stream called?
26. What is a junior right to property?

# 14

# Maps and Plats

## Objectives

After completing this chapter, the student should be able to:

1. Identify the information found on a map.
2. Identify the information found on a plat.
3. Differentiate between:
   A. Topographic and orthometric maps;
   B. Planimetric and profile maps.
4. Describe what is depicted on:
   A. Cadastral maps;
   B. Hydrographic maps;
   C. Isogonic maps.
5. Construct a record of survey map to the standards required for your state.
6. Distinguish between a tentative and a final subdivision map.
7. Relate the requirements of good field notes.
8. Demonstrate proficiency in lettering to industry standards.
9. Demonstrate the proper use of lines to industry standards.
10. Demonstrate proficiency in drafting a map or plat on paper or film in pencil or ink.
11. Demonstrate the proper use of an engineer's scale by plotting a traverse to scale.
12. Plot a legal description, showing bearings and distances to scale.
13. Draft a plan and profile map showing exaggerated vertical scale on the profile.
14. (If available) demonstrate proficiency in the use of CADD by duplicating the drafting skills in the previous objectives using CADD.

A **map** is defined by the American Congress on Surveying and Mapping (ACSM) in *Definitions of Surveying and Associated Terms* as,

> A representation on a plane surface, at an established scale, of the physical features (natural, artificial, or both) of a part or the whole of the earth's surface, by the use of signs and symbols, and with the method of orientation indicated.

A **plat** is defined by the ACSM as,

> A diagram drawn to scale showing all essential data pertaining to the boundaries and subdivisions of a tract of land, as determined by survey or protraction. A map should show all data required for a complete and accurate description of the land which it delineates, including the bearings (or azimuths) and lengths of the boundaries of each subdivision. A plat may constitute a legal description of the land and be used in lieu of a written description.

As a survey technician, you will use maps and plats daily. You must become familiar with the signs and symbols of maps and how to read them as well as how to draft them. As you become more experienced, it will not be necessary to turn a map around to orient it with the ground in order to read it. When you look at a map or plat you will see in your mind's eye a real picture of what a map represents. The contour lines become hills and valleys; the symbols become trees and structures; streams run downhill; swamps bar your passage. With a map and a compass you will never again be disoriented—a surveyor is never lost! The unenlightened will come to you on the job to read the plans for them. Therefore, learn how to do it properly to avoid future embarrassment.

## TYPES OF MAPS

There are many types of maps that you will use or create. **Topographic** maps show the land in three dimensions. The differences in elevation are shown by contour lines. Roads and buildings are shown. Lakes, rivers, and streams are outlined and their depths given.

**Planimetric** maps show the horizontal positions of features as seen from above. **Profile** maps show the vertical differences in elevation as seen from the side.

**Cadastral** maps show the boundaries of the public lands. **Hydrographic** maps show the land beneath the surface of the water. **Isogonic** maps show the lines of magnetic declination and their annual changes.

**Photogrammetric** maps are made from aerial photographs and may be topographical or orthometric, which gives an undistorted view of the surface.

A record of survey plat (ROS) is made to show what has been

done and found in the field pertaining to the lines of ownership of a piece of land. The ROS plat is made to strict standards set forth by the state and is filed with the county surveyor or recorder and mandated by law in some states. A short or long subdivision plat is usually required for a permit to subdivide land.

A **tentative** map is prepared showing proposed layouts of lots and other improvements and submitted to the planning commission for approval. The planning commission will make the necessary changes for approval and return the map to the surveyor. After making the required changes a **final** map is drafted and submitted for approval. When approved, the map is used to stake out the improvements and lot corners for construction.

The simplest map, but perhaps the most important, is the one drawn in the field book to help the office in translating the field notes. A good set of field notes requires no explanation. All the information required to make the plat or map is shown in a form that is easily read by the office technician.

F. William Pafford was a proponent of standardized field notes. He felt that a uniform style, which could be used by the industry throughout the country, was necessary. It would help eliminate the confusion from employer to employer when the technician changed employment. A copy of Pafford's *Handbook of Survey Notekeeping* should be part of every technician's library. The requirements of good field notes are:

- Neatness
- Legibility
- Clarity
- Completeness
- Self-explanatory
- Honesty
- Self checking

The last item, self checking, is the most important. Check your notes before turning them in. A good technician should be capable of helping the party chief by checking the reduction of the field notes each day before returning to the office. There is no worse feeling than waking at 3:00 in the morning knowing in the pit of your stomach that you made a mistake in reducing the field notes and the curb crew will be pouring concrete before you can change the cut sheets. The chance of error is reduced with the increasing use of data collectors. However, the ultimate responsibility is still with the field crew to be sure the information is correct.

When making a field sketch, use the proper equipment. A small scale, a pocket template with various shapes and circles, and a small protractor are the minimum requirements.

**A GOOD SET OF FIELD NOTES**

## A GOOD HAND

You will need a good hand in the field as well as in the office. Practice your lettering, not just until you can do it right but until you cannot do it wrong. Each letter is formed of lines and circles (see Figure 14-1). Practice these shapes but do not try to hurry. Concentrate, instead, on making each line perfect. Speed will come later when your muscles have developed and your hand-eye coordination is synchronized.

Use a lettering guide to make the guide lines so that each letter is the same height. The proper height will depend on the size of

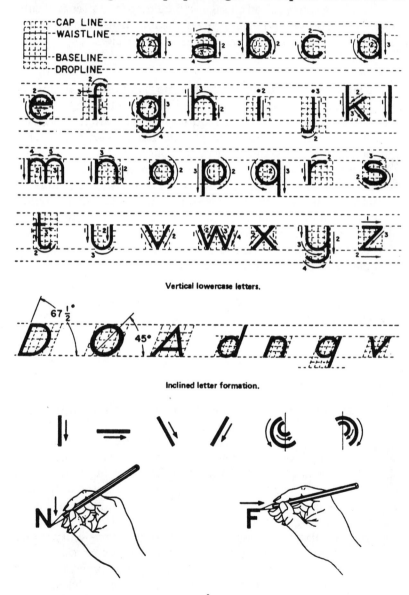

**Figure 14-1** A) Basic lettering; B) Vertical capitals. *U.S. Naval Education and Training Command*

**A**

the space in relation to the rest of the drawing. A ratio of 1 : 2 or 3 : 5 between the upper and lower case letters is preferred by most drafters.

The space between letters should be uniform as measured between the closest parts of the letters. The space between the letters should leave room for an "o." Not all letters are the same width. A, B, K, M, N, R, S, V, W, X, and Z each require a different space. Leave space between the bottom of drop letters g, j, p, q, and y and the next line of letters.

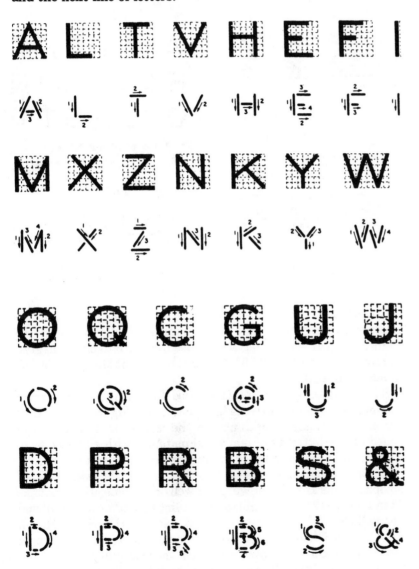

**B**

**Figure 14-2** Numerals and vertical single stroke capitals. *U.S. Naval Education and Training Command*

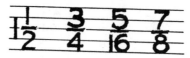

Vertical numerals.

Vertical fractions.

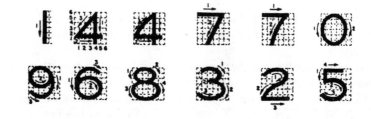

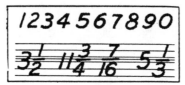

Inclined single-stroke Gothic.

Vertical single-stroke Gothic capitals and numerals.

On numerals (see Figure 14-2), try to use a shape that cannot be mistaken for another if part of the number is lost or smeared (see Figure 14-3). If slanted letters are used, be sure to slant them to the right at an angle of 67.5°. NEVER back slant letters to the left in survey drafting.

Shadow letters are sometimes used for lot numbers or tract numbers to deemphasize the large letter (see Figure 14-4). There are several lettering machines on the market that will greatly speed up the lettering of maps and plats. As with all tools, however, it will require practice to become proficient in their use.

Lines on a map should be uniform in their weight (width) and of an opacity (darkness) that will print well on the blue print machine. Thin lines (.012″ or .30 mm), medium lines (.020″ or .50 mm), and heavy lines (.031″ or .80 mm) are used. Leads of HB, F, H, 2H, 3H, 4H, AND 5H hardness are used for pencil drawings. The proper weight depends on the person's touch, the medium, paper or film, and the temperature. The alphabet of lines (see Figure 14-5) shows some of the common lines used in survey drawings.

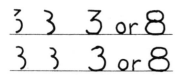

**Figure 14-3** Smeared or obliterated numbers

# SHADOW PRINT

**Figure 14-4** Shadow letters

You will need a good set of technical pens as part of your personal drafting kit. A set with 00, 0, 1, 2, 3, and 4 pens will do most survey drafting projects. Use stainless points on paper or cloth and tungsten carbide points on film. For longer wearing use on film, use jewel points. Be sure to keep your pens clean to prevent clogging of the tip. When reassembling the pen after cleaning the tip, be very careful not to bend the tiny wire on the shaker. When using a technical pen, try to keep the pen nearly vertical to the paper and do not press down. Be sure to use a straight edge that has an undercut side or is held above the paper with inking strips; otherwise the ink may be drawn under the straight edge by capillary action. It is nearly impossible to erase ink on paper, so be careful to plan what you are doing before touching pen to paper. Ink can be erased from drafting film with an ink eraser. If you need to erase anything, use an erasing template to prevent smearing.

If you will be using a drafting machine with a protractor head, make sure it has inking scales. Scales are available in engineering

**Figure 14-5** Alphabet of lines

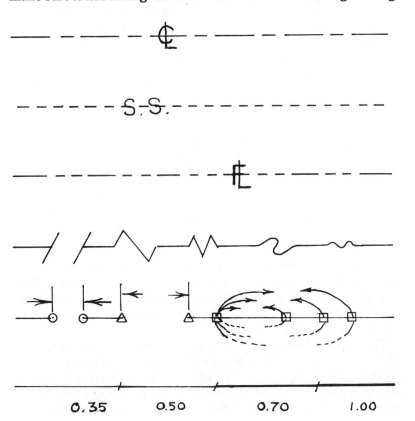

**Figure 14-6** Using a pair of triangles to make parallel lines

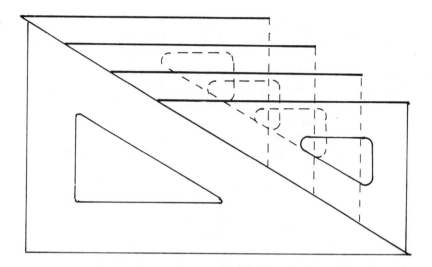

graduations of 10 and 50 or 20 and 40 for use in surveying plans. If you will be using a parallel straightedge, you will need an engineer's scale in 10, 20, 30, 40, 50, and 60 scale. You will also need several triangles ranging in size from 6 in. to as large as you can afford. Get triangles in both 30–60 and 45°. There is also an adjustable triangle that is very handy for other angles. Triangles can be used in pairs to make parallel lines (see Figure 14-6). Get a good circular protractor that is graduated in both directions. Some protractors are graduated to 360° on one scale and 0 to 90° quadrants on the other scale. They are very useful for plotting bearings.

A good 6-in. compass with a pen adapter and an extension bar is useful; but most drafters would rather use circle templates in many cases. Get a circle template that has civil engineer's scale rings and several sizes of round holes for survey drafting. Several shapes of French curves can be used for drawing the curved lines used in vertical curves and road profiles. A kneadable eraser is good for cleaning up the final drawing and a desk brush is necessary to keep your work area clean.

## THE BEGINNING DRAFTER

The beginning drafter always spends too much money on gimmicks that are nice but rarely used. Look around a survey office. See what drafting tools are really used and buy those. Always buy the best quality tools you can afford. They will more than pay for themselves over the years.

The simplest drawing is plotting a **traverse**. A traverse is a series of angles and distances that run between two points or close back on the point of beginning. The first method of plotting is using a protractor and a scale. If the field angles are given, the first line to be drawn will be the basis of bearing (BOB). Start with the point of beginning (POB) and draw a line through the point along the BOB. Place the protractor on the drawing with the center point on the POB and the index line (0°) on the BOB line. Count off the proper number of degrees from the backsight to the next point. Angles may be right, left, or deflection. Make a point with your pencil on the proper degree point and leave the pencil point on the paper. Place the zero mark of the scale on the POB and the edge of the scale against the point of the pencil (see Figure 14-7).

Move the pencil point to the scale distance and make a point. Using the proper symbol, put a symbol around the point and number it. Using a straight edge draw a line between the two points. Move the protractor to the new point and align the index line along the backsight line to lay off the new angle. Continue in the same way through all the traverse points to the last station. If it is a connecting traverse, plot the last point and lay off the closing angle with the protractor. If the traverse closes on itself, there will probably be an error in the closing line. The error is usually caused by not being able to lay off the angles exactly and not by a blunder in the plotting.

A more accurate method of plotting the angle points is making

**Figure 14-7** Using a protractor and scale to plot a point

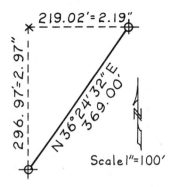

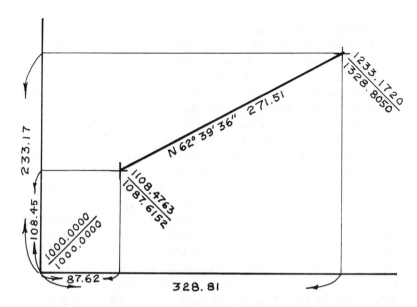

**Figure 14-8 (Above)** Using latitude and departure to plot a point

**Figure 14-9 (Right)** Using coordinates to plot a point

use of the trigonometry you learned. Extend the line through the second point a distance equal to the distance calculated by using the sine of the deflection angle times the distance between the points. Construct a right angle to the extended line and measure over the tangent distance. Mark the point with the proper symbol and continue around the traverse as before (see Figure 14-8). The closing error should be less.

The most accurate method of plotting points is by using the rectangular coordinates of the points. Start with the POB and, using the parallel arm, lay off the easting of the second point. Next, using either a large triangle or constructing a right angle to that point, lay off the northing (Ng) of the second point. Continue to plot each of the traverse points by using their adjusted coordinates. Finally, connect the points with lines. The closure will be automatically correct (see Figure 14-9).

## THE PLOTTING OF LEGAL DESCRIPTIONS

The plotting of legal descriptions of lots and easements is similar to plotting of traverses. Start with the POB and plot each course as called around the description. Label each course with the bearing and distance and identify the corner monuments (if called on the description or found in the field notes). If details are to be plotted inside the boundary, it is best to place the bearing and distance outside the property lines. Plan what else is to be put on the drawing before doing any lettering to avoid having to erase and move it to another position.

Room must be left for the surveyor's certificate and the auditor's certificate on record of survey maps. The legal description may either be lettered on the map by hand, by machine, or typed on another transparent sheet and affixed to the drawing with clear tape. The north arrow should point to the top or left hand side of the sheet. This way, when the map is bound in a book at the recorders' office it may be read from the bottom or right hand side.

The standard size for a map to be recorded is 18 by 24 inches. Draw a .5-inch border on the top, right, and bottom. Put a 2-inch border on the left to allow for binding. Most states require two legible prints of each survey on a permanent type of medium (tracing cloth or equivalent) using black ink. The lettering must be of a size and quality that will allow it to be photographically reproduced and reprinted.

## A PLAT MAP

A "plat map" must be filed for all "short" or "long" subdivisions. A short subdivision may be defined as "The division of the land into four (check with your own state laws to determine the number of lots allowed on a short subdivision) or fewer lots, tracts, parcels, sites, or divisions for the purpose of sale, lease, or transfer of ownership." Short plats usually have different requirements than those for a long plat. They normally do not allow for the land to be again divided, for a period of time, without the filing of a long plat. This prevents people from getting around the platting laws by resubdividing the short plat. A long subdivision is a lengthy and involved process used by the state to determine if appropriate provisions have been made for open spaces, drainage, streets, water, sanitation, schools, police, fire protection, and any other relevant facts.

A tentative or preliminary map is required to be submitted for approval by the governing body. The tentative map will show the boundary and topography of the land to be developed. The proposed layouts of lots, streets, walks, utilities, sewer and storm drains, as well as existing improvements are also shown. The name of the subdivider, the name of the engineer or surveyor, the legal descriptions of the land, the tract name and number, and a location map are placed around the map. It is very important that all the required information is shown in a legible way. The members of the planning commission may not be trained in map reading so the map should be self-explanatory to avoid delays in approval caused by lengthy questioning of its content.

## PLAN AND PROFILE MAPS

When tentative approval has been given, the streets are laid out and profiles and cross sections are run for design purposes. Plan and profile maps are made from the field notes. Streets, curbs, sidewalks and underground utilities are designed and shown on the maps. Plan and profile maps for sanitary and storm sewers are also prepared from the field notes, showing the location of existing sewers and the inverts of the nearest sewer manholes. If the existing sewers are off the map, a map of the proposed connecting lines must be made and easements obtained.

## A GRADING PLAN

A **grading plan** must be made for the lots, showing the drainage to the street or alley and the elevation of the house subgrade. The earthwork quantities will be taken off the plans and total cut and fill amounts calculated. The cuts and fills are calculated from cross-section drawings of the streets and finish contour maps of the lots.

## A TENTATIVE TRACT OF A SUBDIVISION

Using the adjusted boundary from chapter 7 and the topography drawn from chapter 8, draw a tentative tract map of a subdivision. The lots are to have a minimum of 15,000 square feet and a maximum of 25,000 square feet. The lots will have a minimum of 100-foot frontage on regular lots and 60 feet on flag lots (a flag-shaped lot, placed behind the lot bordering a street, with the "pole" section being an access easement). Streets will be 50-feet wide (see typical section) with a 25-foot curb radius on corners and a 50-foot radius on cul-de-sacs. All lots will be served by underground utilities. Cut slopes will be 1 : 1 and fill slopes will be 2 : 1. An existing street runs along the east property line. The street is 60 feet wide and drops at a constant 3% grade from north to south. An existing sanitary sewer made of 24-inch vitrefied clay pipe (VCP) runs 5 feet east of center line. A 10-inch water line is 10 feet west of center line. A 36-inch storm sewer is 15 feet west of center line. An existing natural gas line is located 15 feet east of center line. Telephone and cable television conduit is buried 2 feet from the east property line toward the street. The center line profile notes for the street show that 0+00 was taken opposite the south property corner. The stations and elevations are

$$0-100 = 96.72$$
$$0+00 = 99.70$$
$$0+40 = 101.00$$

At 0+40, 15-feet right of center line is a sanitary sewer manhole.

$$
\begin{aligned}
\text{rim elevation} &= 100.90 \\
\text{invert south outlet} &= 92.85 \\
\text{invert north inlet} &= 92.95 \\
1+00 &= 102.70 \\
3+00 &= 108.70 \\
4+00 &= 111.70
\end{aligned}
$$

$$5+00 = 114.70$$
$$6+00 = 117.70$$
$$6+40 = 124.90$$

At 6+40, 15-feet right is a SSMH

$$\text{rim elev.} = 118.90$$
$$\text{outlet invert south} = 110.85$$
$$\text{invert north} = 110.85$$
$$7+00 = 120.70$$
$$8+00 = 123.70$$
$$8+40 = 124.90$$

opposite the north property line, 9+00 = 126.70. The street has a 2% cross fall with 20′ traffic lanes (18 feet each side of center line). There is a 6-inch concrete curb with a 2-foot concrete gutter on both sides of the street. The parkway is 10 feet and there is no sidewalk. There are storm drains catch basins at 1+50, 3+50, 5+50, 7+50, and 9+50 in the gutter on both sides of the street. The depth of the storm drain is to be determined in the field.

The first step is to draw the profile of the existing street and curb, then show the existing utilities. Draw a typical section of the street and put in the underground utilities. The storm sewer should be higher than the sanitary sewer. The other utilities should be 2- to 4-feet deep. These would normally be located on the surface by the utility company and the depth would be found by digging. BE SURE TO CALL THE UTILITY COMPANIES BEFORE DIGGING IN THE FIELD IF YOU KNOW THAT THERE ARE BURIED UTILITY LINES. They are not always where they are shown to be on the plans.

Design the entrance road or roads to your subdivision and calculate the elevation at the center line intersections. Try to keep the street grades at least 2% to allow for drainage of surface water along the gutters. Make a rough sketch of the streets and try to fit in the lots as you go. Move the streets to allow for lot size. When laying out lots, try to imagine how they would look if you bought one. Work with the natural contours of the ground to avoid having to move any more dirt than is necessary. Try to balance the cut and fill on the lots to maintain drainage to the street at 2% or more.

Do not design driveways with more than a 12% grade and plan for clearance in sags and humps or summits. Sewer connections should be at least 2-feet deep at property line and join the sewer main with a "Y" connection flowing downhill.

A lamp hole is needed to any dead ends and a manhole at all angle points in the sewer. All other utilities are stubbed out to property line. Water, gas, and telephone lines are usually run to the sideline of every other lot and "y'd" to each lot with a meter and shutoff valve. Storm drains should be placed so as to intercept the water running down the gutters before it runs across the roadway. They must also be at intervals along long stretches of curb. The distance will depend on the run-off and will be calculated by an engineer from the average rainfall for the area.

An engineer will calculate the size of the sewer pipes based on the number of houses on the line. The sewer pipes usually start at the uphill end with a 4- or 6-inch size line and get progressively larger as more houses are serviced by the line. Sewer manholes are 4 feet in diameter and have a .1-foot cross fall between inlet and outlet. The distance from the rim elevation to the flow line is called the invert.

For the road, sewer plans, and profile maps, use a sheet of specially made plan-and-profile paper. Fasten the paper to the drafting table with the clear section to the top. The orange profile pattern will be on the back, or toward the desk. The printing on the paper side will be seen through the paper to show on a blueprint but not interfere with drawing.

Use a scale that looks right for the size of the sheet and shows the necessary detail. Because of the low relief of the vertical distances, an exaggerated scale is used for the profile as opposed to the plan view. If the horizontal scale is $1'' = 40'$, then the vertical scale should be $1'' = 4'$. The vertical lines on the profile section should line up with the stations on the plan view and are spaced each 10 feet at 40 scale. The horizontal lines are in groups of five for .5 feet on the dark lines and .1 ft. for the light lines. A title block is usually printed in the lower right hand corner for the company and shows the drawing number and scale for your drawing. Just make one up if your school does not have one.

The plan portion must be parallel with ruled lines for the stations to line up vertically. If there is a curve or angle point in the road or sewer, a break must be made in the plan to remain parallel. If the profile view gets too close to the top or bottom of the ruled area, a vertical break is made by drawing a vertical line through the point and sliding the elevation up or down the line to a more usable point (see Figure 8-17). The percentage of grade for the road is shown along the profile between vertical curves. The percentage of grade plus type and length of pipe is shown along the pipe profile between manholes.

Cross-section drawings are also made on special paper. The most convenient type is a 10 grid that allows the counting of squares for area. A special cross-section paper is also made that has an exaggerated vertical grid the same as profile paper. This requires more care in plotting but allows the vertical differences to be more easily seen. If you are going to calculate the area using the grid coordinates, you can use the cross-section paper. If you are going to count squares or use a planimeter, the 10 grid will be easier.

The "grading plan" for the lots is a plan showing the finished grades of lots superimposed over the topo contours. The finish-grade contours are plotted and the areas of the difference between the finished grade and the original ground is found using a planimeter. The areas found are multiplied by the vertical distance between the contours to find the volume of each level. This is similar to the average end area computation, only the end areas are horizontal and the distances are vertical. The areas of cut are calculated separately from the areas of fill and the totals of each are given. The difference between the cut and fill quantities is the amount of export or import material needed.

The "final map" will be made after all of the changes have been made and the roads and utilities have been designed and placed in their final location. In some cases the final map will reflect the as-built survey information. A title sheet will have the necessary certificates and signatures. When the final map is approved and recorded, the street dedications become public property. The lots are recorded simultaneously.

---

Now that you have learned what needs to be put into a map and how to do it the hard way, it would be nice if you could get a machine that would do all that hard work for you. The answer is "computer-aided drafting and design," CADD for short. There are several fine surveying and civil engineering CADD packages on the market today. Some packages are dedicated stand-alone systems like the ESP-200 (Engineering Survey Package) by PacSoft Incorporated, of Kirkland, Washington. Other packages operate with a drafting package such as AutoCAD for the CADD system and write specialty programs for surveying to run on it.

A knowledge of computer programming is not necessary to run the CADD programs. All the programming has already been done for you. A general knowledge of how a computer works is all you need. A working knowledge of the DOS system used by the computer will be helpful if you run into a problem in the I/O (input/output) system.

The PacSoft ESP-200 does not require much training. A walkthrough practice problem teaches the fundamental functions and how they are used in a stair-step method with each new skill leading to another. The first volume is COGO-200, which allows the field information to be entered into the computer either by hand or by interfacing with a data collector.

The field angles and distances are reduced to northing, easting, and elevation (X, Y, Z) and can be given a descriptor. With the "in-

**COMPUTER-AIDED DRAFTING AND DESIGN**

stant image" feature turned on, each point is shown on the screen as it is calculated. The points are put into "sets." The commands can be saved under "command memory" for use in editing or correcting data and, later, for recomputing area, volume, or other data. Points may also be entered using a digitizer. This allows the input of points from another map or drawing and rescaling to the scale of map you are making.

For traverses, you can balance the angles and adjust the elevations using internal or external reference points. You can choose between compass rule, transit rule, Crandall's, or least squares adjustments. You can do lines, curves, inverses, bearing-bearing, bearing-distance, and distance-distance intersections.

There are eighteen different kinds of curve design options in the curve menu. They include curve passing through 3 points, curve passing through 2 points, tangent to a line, curve with known radius passing through 1 line, tangent to 1 line, etc. For laying out lots, you can use a predetermined area to solve for 2 points, closing off a figure using a sliding side, a hinged side, or radial to a curved side.

Use the figures and alignments menu to compute up to 12 points. Give the center line of two intersecting roads, the road widths (they need not be the same), and the curb return radius or tangent lengths.

The program works on straight or curved streets or cul-de-sacs. You can generate reports for area/lot summary, radial stakeout, baseline staking, horizontal alignment/curve staking, and offsets to a line.

Section subdivisions are done on a special menu that will do lost corners by single or double proportionate measurement and will correctly handle the north and west sections of a township.

The coordinate conversion menu will do plane to **ellipsoid**, ellipsoid to plane, and plane to plane in adjacent zones' conversions and also state plane coordinate computations using lambert, transverse mercator, and oblique mercator projections for all 50 states and the American possessions.

The spirals menu does transition spirals and circular curve computations based on the Euler spiral. Spirals can be stored as sets to allow you to compute spiral intersections, spiral offsets, and areas bounded by a spiral. With the command memory feature, you can change the width of a street and have all the adjoining lots corrected automatically.

The PacDraft-200 volume uses the coordinate files from COGO-200, ROADS-200, or DTM-9000, which is PacSoft's digital terrain modeling system, to make finished drawings. The drawings can be customized by using the user-definable keys to make title blocks, company logos, and your own north arrows.

The keys menu allows you to choose the line type, the point marks, line and curve annotation, point labels, and annotation identifiers. There is also a TEXT EDITOR that allows you to write legal descriptions, going point to point, just by typing "1*2." You will get the bearing and distance between points 1 and 2 in the description.

The "drafting sequence file" holds the point numbers defining the line or curve, the line type key, the symbol key, and the rotation of the symbol. It also holds the line/curve annotation key and the location of the annotation, right, left, or center justification, the point label, and a comment up to sixteen characters. The drafting standards are used to select formats and options for all final plats. The choices are character slant, right justified text, direction arrows, feet or meter marks, paper size, paper orientation, and paper roll for continuous plotting of spooled files.

The "plot generation file" allows you to choose the border and title block location and plot layout. This allows the placing of map, text, symbols, and tables just by pointing to the position with the cursor and entering. Layers can be plotted separately to make different types of drawings from the same sequence file. When plotting, if there is not enough room to place the line annotation, a line or curve table will be created and the information entered and numbered in sequence.

The third volume for the ESP-200 package is ROADS-200. It deals with roadway and highway design, earthwork and volumes, plan and profile sheet drafting, cross-section plotting, and various other drawings and reports common to roadway engineering. The data can be input from COGO or directly input into ROADS, as there is a subset of volume 1 routines that can be used.

The "vertical control module" can be used to define the original ground profile and the final grade profile. It is possible to use an exaggerated vertical scale and the instant image gives an ongoing picture of the vertical alignment along with stations and elevations.

Profiles can be input directly from field notes or from the keyboard in the form of station and elevation, station and percentage of grade, or from a back and ahead station with a grade from each to a VPI. The profile can be created from cross sections taken on another line. You can also enter a profile by using the digitizer. A vertical curve can be entered by inputting the curve length, whether symmetrical or asymmetrical, by station and elevation or the station of the high or low point.

Cross sections to the profile can be entered as offset and elevation pairs, up to 200 of them, and perpendicular or skewed to the control line. A descriptor can be entered for each point such as top, toe, or EP (edge of paving). Sections can be realigned along a new alignment if the final alignment is changed and a skewed section straightened to help in volume computations. Cross-section templates can be defined for any surface (finish grade or subgrade). Simply define the distance and vertical change for the grade change points or nodes. The program can look for catch points outside the right of way and can change the slope to fit and flag the point with a warning message.

For transitions in super elevation or change of roadway width, only the beginning and ending templates need to be entered. The intermediate templates are automatically created. The design of each element of the roadway is stored separately. The design file is used to assemble them into a design surface, which allows "what-if" chances without having to reenter all the points affected.

Up to nine layers of material can be handled under volumes. Shrink and swell factors can be applied to each layer. Either average end area or the prismoidal methods can be used. It is possible to change the grades and see what effect this has on the quantities just by moving the VPI. Printouts include the areas of cut and fill, bank volumes of cut and fill, loose volumes of cut and fill, and the mass ordinate of each station.

The "reports" section allows the following choices:

1. Layout from a baseline for setting grade stakes and offsets from a specified base line.
2. Radial stakeout for setting grade stakes from specified control points.
3. Slope staking for setting the catch points for slopes.
4. A cut sheet summarizes the cuts and fills from the stake out hubs set in the field.
5. Linear quantity allows the computation of the length of curb and gutter, guard rails, or whatever along the surface of the ground rather than by station.
6. Surface area computes the surface area quantities for sidewalks, parking lots, pavement, hydroseeding, etc. These are slope areas.
7. Volumes and mass haul give the ability to define how various materials contribute to the volumes and mass diagram and to figure the shortest haul routes.

Up to four additional utility sets can be included in each cross section. The system will calculate the proper position and plot the symbol for each utility at its proper position horizontally and vertically. Plan-and-profile sheets can be drawn on plain paper or on preprinted grid paper. Profiles can be plotted at the top or bottom of the sheet in any proportion of the total sheet.

The most amazing feature of ROADS-200 is the ability to do a perspective drawing of the roadway as seen from any selected station along the alignment. Hidden lines are removed and the

**Figure 14-10** ROADS-200. *Courtesy of PacSoft*

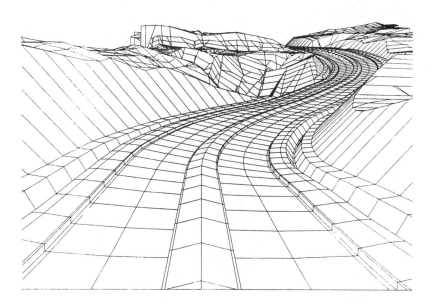

output may be seen on the monitor or drawn on the plotter. This allows the designer and the client to see what the finished roadway will look like to the driver of a car traveling along the road in either direction (see Figure 14-10).

For topographical mapping, the digital terrain modeling or DTM package by PacSoft is state of the art (see Figure 14-11). The T-net modeling and mapping portion of the DTM system uses a triangular irregular network of the original points. T-net allows you to define natural break lines that will be used by the system to decide which points to include in the triangle. Tops, toes, grade breaks, as well as flow lines of creeks can be marked and labeled. The boundary can be trimmed or extended. Points can be switched to fine tune the contours without having to create an entire new model. The instant image allows the work to be seen as it progresses so that changes can be made before printing.

The topographical mapping systems will produce complete topographic maps including borders and title blocks. Notes, planimetric features, reference grids, symbols, and labels are all included. The maps can be reversed for plotting on the back of transparent media. Large maps can be drawn on several different sheets with the edges butt-matched or overlapped. Intervals, pens, colors, thickness, and pattern can be selected for major and minor contours. Labels can be drawn at breaks along the contour line selected by the user. Minor contours can be dropped in very steep terrain as specified by the operator.

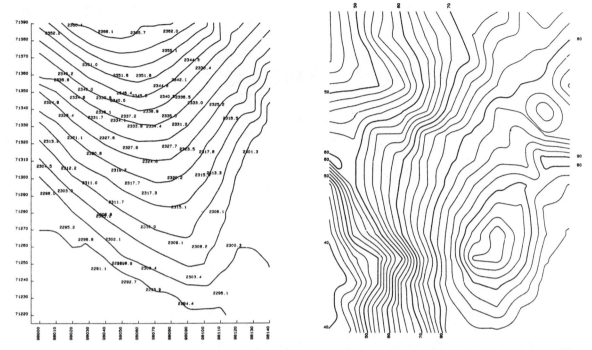

**Figure 14-11** Contour mapping.
*Courtesy of PacSoft*

You may choose splines, no smoothing, or circular curves for contour smoothing between points. Points can be labeled by set, point number, or descriptor, such as connecting all points named "curb" or "flow line." Symbols can be placed by digitizing the position at point numbers, at coordinates, and at set points. Symbols may also be placed at selected descriptors, such as "fire hydrant" or at even intervals along a set.

The second portion of the DTM package covers "section modeling and mapping." Ten separate ground or material surfaces can be specified and plotted at each station. These designations allow easy recognition of cut and fill or changes in the type of material to be moved or placed. Up to ten utilities can be shown and identified at each section. With the utilities and ground surfaces shown to scale, it is easy to check clearance between water and sewer lines or to change grade to assure that there is enough cover over water and gas lines. The cost of broken telephone-line repair can run into thousands of dollars for interrupted service. If the telephone lines are shown on the cross sections at an accurate depth and location, the excavator can take steps to assure the safety of the utility.

The third portion of the DTM package is "3-D mapping" (see Figure 14-12). The ability to show the proposed project in a three-dimensional view will allow the technician to spot design errors more readily than in a two-dimensional drawing. It will also be easier for the client who is not familiar with reading maps to see what has been proposed and make suggestions for changes.

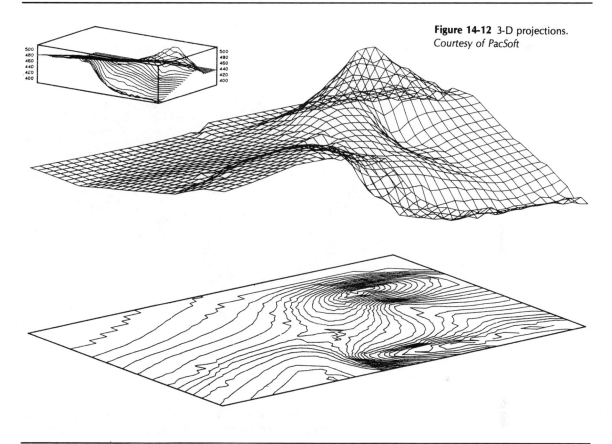

**Figure 14-12** 3-D projections. *Courtesy of PacSoft*

## THREE-DIMENSIONAL PROJECTIONS

Three-dimensional projections allow the terrain to be viewed as if the viewer were in an airplane flying over the site. The rotation and tilt can be changed to present a different view point of the terrain. Three-dimensional perspectives allow the viewer to see what the site looks like to an observer on the ground looking in any direction. Normal, wide angle, and telephoto views can be shown.

Two-dimensional figures can be superimposed on the three-dimensional surface to show building sites, roads, and grading (see Figure 14-13). The shading from a light source can be shown to give a realistic effect to the scene. Lines that would not be seen (hidden lines) are automatically removed to avoid confusion. The surface can be shown as a one-way mesh or section plot. The surface can also be shown as a two-way mesh or fishnet that allows the changes in terrain to be seen with greater clarity. Contours can also be shown in projection or perspective views.

Another PacSoft program, CADLINK-9000 allows the ESP-200 and DTM-9000 data to be converted into the graphics standard format. HPGL (spooled plot) files can be translated to DXF or IGES files used by AutoCAD, VersaCAD, and others with this package. PacSoft coordinates can also be translated to DXF or

**Figure 14-13** Fishnet overlay.
*Courtesy of PacSoft*

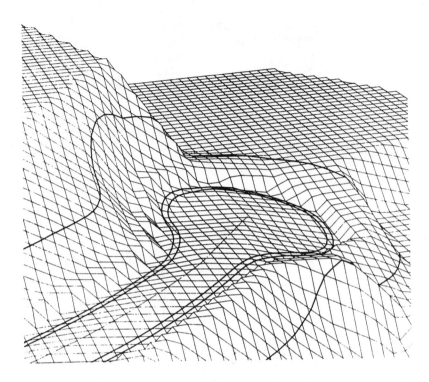

IGES format and vice versa. This will allow the direct transfer of data and drawing files between different offices and even different companies.

While this may all sound overwhelming to you, it is really quite easy to learn CADD if you have a good drafting background and are able to follow written directions. ESP-200 is but one of many systems on the market and although it may not be the one you will be using, the principles will be the same. Remember, CADD does not think. You are the thinking part of the system; the computer will only help you do your job. Do not be afraid to learn new systems. As you gain experience and expertise, you will be a more valuable asset to your employer and achieve job security and advancement to more responsible positions.

## Review Questions

1. Define a map.
2. Define a plat.
3. Name five types of maps.
4. What type of map is used to show the proposed layout of a subdivision?

5. After the planning commission has made changes to the map in question 4, what kind of map is then prepared for use in staking out the improvements and lot corners?

6. List the requirements for good field notes.

7. On a sheet of paper, using pencil and in your best hand, do an upper case and lower case alphabet and numerals 0 to 9.

8. Following the instructions on pages 252 through 255, make the tentative map of a subdivision. This map should be in ink on vellum. The original plus a reproduction will be given to the instructor on completion.

# Survey Data Systems

**15**

## Objectives

After completing this chapter, the student should be able to:

1. Identify and compare some of the total station survey system equipment and software currently in use.
2. Define Geographic Information System and relate how the system may be used in industry.
3. Define Land Information System and relate how the system will help with land ownership research in industry.
4. Recognize and cite evidence regarding: the need for positional tolerance in the location of boundary corners under GIS/LIS systems.

For centuries surveyors have gathered data about the land. These data are used to locate and divide the land and to make improvements. Recent changes in technology have made the gathering and recording of data faster and easier for the survey technician. The data, when collected, can be transmitted to the office with fewer chances of error. Once in the office, the technician can manipulate the data with greater speed than ever before possible through the use of computers. The data can be easily stored and retrieved without having to use huge file cabinets and a stack of maps. Once the design work is completed, the layout data are taken to the field and the points can be set with little field time used for calculation. Many points can be set from one point without moving the setup, saving field time. The data, once reduced to usable form, are used in Geographic Information Systems (GIS) and Land Information Systems (LIS) to make the information available to many agencies and user groups.

## THE TOTAL STATION SURVEY SYSTEM

The total station survey system (TSSS) has made much of this possible. The TSSS consists of a total station with a data collector, a field computer, and a CADD station in the office.

In honor of the 100th anniversary of the founding of the Carl Zeiss Foundation, we will use the Zeiss Elta 2 total station (see Figure 15-1) as the instrument. We will also use the Zeiss REC 500 Electronic Fieldbook (see Figure 15-2) as the data collector and the Hewlett-Packard Vectra QS/16 computer with PacSoft ESP-200 software (see Figure 15-3).

There are many other combinations on the market that will give excellent service but covering just one combination should be less confusing to the technician.

The Zeiss Elta 2 series E is nominally accurate to 0.6″ of horizontally and vertically. The EDM is electro-optical with modulated infrared light coaxial in the theodolite telescope. It is accurate to ± 2 mm ± 2 ppm. This information is needed to figure the positional tolerance of the instrument. All measuring and computing processes are controlled with three keys for operator convenience. The four LC digital displays on the front and rear of the instrument prompt the operator through the programs and display the meas-

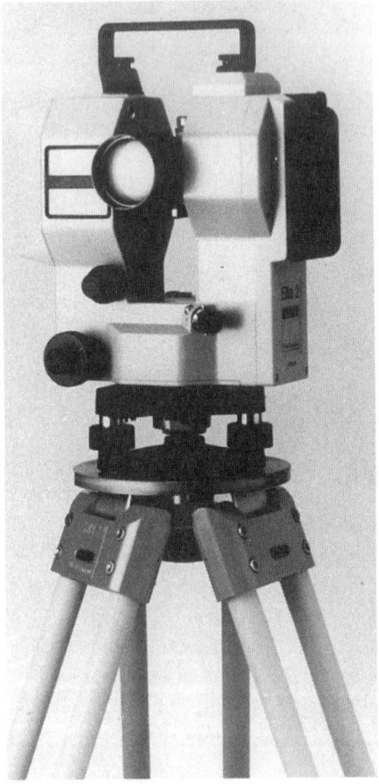

**Figure 15-1** Zeiss Elta 2. *Courtesy of Carl Zeiss, Oberkochen, Germany*

**Figure 15-2** Zeiss Rec 500 Intelligent Field computer. *Courtesy of Carl Zeiss, Oberkochen, Germany*

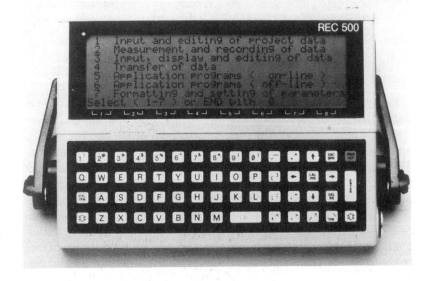

**Figure 15-3** Hewlett-Packard Vectra QS/16 computer. *Courtesy of Hewlett-Packard Company*

urement results. There is an automatic correction of error sources such as vertical axis inclination, circle eccentricities, culmination and index errors, and trunnion axis error. The display reads to 0.6″ and the maximum distance is 3600 m. The following programs are called by pressing a key:

Reduction of distance
Input of station coordinates and target point eccentricities
Determination of polar coordinates in the national system or a local system
Determination of check distances in a random combination
Determination of object heights
Determination of point distances in relation to a definable line
Setting out with continuous display of distance and offset from the trial point to the point to be set

The Zeiss Rec 500 data collector enables automatic and manual data acquisition, storage, and editing. Data is transferred to and from the unit by the RS-232 integrated interface. The different measuring and computation modes and the necessary inputs are shown on the display. The unit is user friendly, guiding the operator through the necessary measuring and computation processes in the dialog mode. It even uses graphics in some programs.

The system uses twenty-seven alphanumeric and numeric characters for point numbering and identification. The point it identifies is stored with the data. One data line can contain eighty characters. The date, project name, party, barometric pressure and temperature, etc. are input using a regular qwerty (standard typewriter) keyboard.

Data processing can be done both on and off the line, during or after the measurement. A maximum of 2000 data lines can be stored. Intermediate data are stored on a battery powered 3½"-disk drive accessory. Data can be sent and received on two software controlled interfaces and additional optional computer interfaces. The total capacity of the REC 500 is 352 kB, allowing approximately 130 kB of special programming and data storage by the user. The operating system is CP/M.

## RAW FIELD DATA

After work in the field, the data collector is taken into the office and downloaded into the office computer through the RS-232 interface. The data collection portion of the PacSoft ESP-200 package allows reading the data directly into the computer, including point codes, descriptors, etc.

Once in memory, the raw field data can be listed, edited, reduced, and stored onto a disc file. A hard copy of the raw field data can be run. The field data are reduced directly into a coordinate file that can be used by PacSoft's ESP-200 and DTM-9000 packages. The use of numbered text codes allows the storage of up to 250 alphanumeric character strings on the program disc. These can be used for comments during field input, party names, or descriptors, which will be stored with the solved points. The data collector records only the text index number.

When the data are reduced, the computer will look up the corresponding data in the text index file and print the stored text string. The compatibility of the REC 500 and ESP-200 allows the REC 500 to be used as both a data collector and a field computer in this TSSS. The use of a lap-top computer for field editing may be necessary in some other systems.

## THE HAND-HELD CALCULATOR

In 1972, Hewlett-Packard pioneered the era of personal computing that is a standard of the industry today. The HP-35, the first scientific hand-held calculator, and the programmable HP-41CV, still in use by many surveyors, soon made the engineer's slide rule obsolete (see Figures 15-4 and 15-5).

**Figure 15-4 (Above)** The HP-35 calculator. *Courtesy of Hewlett-Packard Company*

**Figure 15-5 (Left)** The HP-41CV calculator. *Courtesy of Hewlett-Packard Company*

## A NEW GENERATION OF CALCULATOR/COMPUTER

A new generation of calculator/computer began with the HP-28S scientific and the HP-48SX (see Figure 15-6). The HP-48SX has replaced the HP-41 as the newest in a long line of Hewlett-Packard hand-held calculators. Many systems use the HP-48SX as a data collector (see Figure 15-7). It is an understatement of facts to call the HP-48SX a calculator. There is 256 KB of ROM and 2100 functions built in. There is 32 K of RAM expandable with add-on cards to add 128 K RAM and still have room for a survey card. One of the survey cards already out for the 48SX is produced by Tripod Data Systems and can handle up to 3000 points. The HP-48SX can be hooked up with a serial cable directly to the total station and used as a data collector. By adding an RS-232 adapter to the cable, you can download directly into the office computer.

CMT Incorporated, of Corvallis, Oregon, has introduced the MC-II and MC-V (see Figure 15-8). The MC-II and MC-V are actually hand-held computers designed for applications and environments where true portability is required. The MC-V is the environmental version of the MC-II and is probably the most rugged hand-held computer ever built. A built-in heater allows operation in temperatures as low as −40°F. The watertight, dust-proof case allows the MC-V to operate even when submerged in water. The MC-V has two built-in RS-232 interfaces to allow communication with the office computer. There is up to one megabyte of direct memory

**Figure 15-6** The HP-28S calculator. *Courtesy of Hewlett-Packard Company*

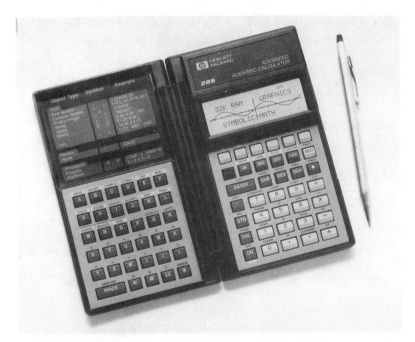

**Figure 15-7** The HP-48SX calculator. *Courtesy of Hewlett-Packard Company*

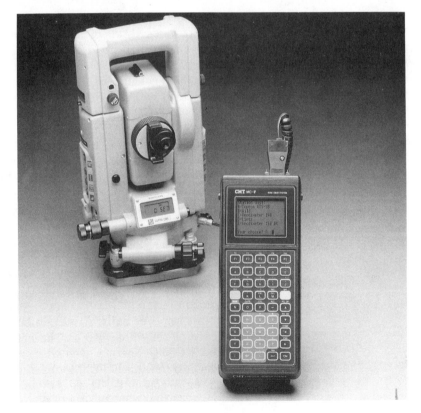

**Figure 15-8** The CMT MC-V Field computer. *Courtesy of CMT, Inc., Corvallis, Oregon*

addresses in each computer. Memory is expanded by plug-in ROM on battery-backed RAM. An optional RAM disc holds up to 1 mB of mass storage.

There is an HP-41 emulator that allows the MC-V to run application packages developed for the HP-41CX. The MC computers also run MC-BASIC (GW-BASIC compatible), MC-MBASE (dBASE III plus compatible), MC-SPREADSHEET (LOTUS 1-2-3 compatible), and Microsoft "C". The programs are available as plug-in modules. The MC-II Surveying Pac contains all the HP-41 CV/CX programs published by D'Zign for land surveying. The MC can be connected to several different total stations, by an optional cable, for use as a data collector.

## ESTABLISHMENT OF A GIS/LIS SYSTEM

All of this automation helps to speed up the collection and recording of field data but does not relieve the technician of the need for training. The need for the technician to upgrade his or her training increases as surveying techniques change. Some survey firms are moving in the direction of land-use planning. This will require the technician to become familiar with the other aspects of land ownership and use. The need for faster and easier access to land records has led to the establishment of a GIS/LIS system in many states.

The ACSM publication *Multipurpose Cadastre Terms and Definitions* (available from the ACSM) defines the Geographic Information System as, "a system of hardware, software, data, people, organizations, and institutional arrangements for collecting, storing, analyzing, and disseminating information about areas of the earth." A Land Information System is defined as "A geographic information system having, as its main focus, data concerning land records."

The GIS/LIS system must begin with a base map that must have three features:

1. Precise geographic location such as state plane coordinates on each feature;
2. The relative position of each feature relative to the other features on the map; and
3. Descriptive data that will provide pertinent information about the feature such as ownership, past and present, encumbrances, utilities, etc.

The precise location of features such as administrative boundaries, block lines, buildings, easements, public utilities, land boundaries, physical geography, topography, and zoning will differ depending on relative importance of each feature. Some features will need a high order of precision in location, whereas others such as drainage areas, will not. Locations may be found by use of GPS or ISS for accuracies in the order of 1 : 100,000 for high precision features. Aerotriangulation can be used for second-order location of secondary features. Digitizing from existing maps of proven accuracy can be used for large areas requiring lower precision. Orthophotography can be used for location of photo-identifiable features using a digitizer. A scanning digitizer will cut down on the time for entry of existing points into the data base.

Relative position or positional tolerance required for each point will also vary according to the importance of exact location for that feature. Survey technique and equipment will need to be planned to meet the positional tolerance required for each point of the survey. Higher precision will be needed for the control monuments than for street center-line monuments. Lot corners will need to be more precise than building corners. The location of public utilities, both above and below ground, will need to be measured with more precision than the boundaries of the utility easements.

The descriptive data will allow the map maker to make a decision about which layer of the database to use for the feature. Some of the layers might be:

Topography
Zoning
Hydrography
Ownership
Street networks
Fire control districts
Water mains
Wetlands
Soil type

As the GIS/LIS systems become larger, they will meet other systems started by adjoining municipalities. To assure that the systems will be compatible, a nationwide standard must be implemented. The use of state plane coordinates tied to NAD 83 will make a national system possible. Computation of least square adjustments and state plane coordinates on lot corners will become more common with the use of computers. If land is recorded by a parcel identifier and the corners have state plane coordinates, a computer search for adjoiners will reveal any possible conflicts of location.

Just a few years ago, a couple of feet more or less did not make much difference on rural property. Today, the need to eliminate conflicts between adjoiners increases as land becomes more valuable. Rural property is rapidly changing to urban home sites as the need for more housing grows.

Many of the surveys made have not been recorded owing to the lack of a recording act. Although some states now have acts that require a survey to be recorded and a plat filed, previously surveyed lands are not shown. As the unrecorded surveys are discovered and resurveyed, they must be added to the GIS/LIS. This need for new surveys and resurveys will create a demand for technicians that are familiar with up-to-date methods and equipment. Trained technicians will be needed to enter the information into the data banks as the surveys are completed and recorded. The future for trained and certified technicians looks very bright at this time. To change from the present recording system to a nationwide GIS/LIS system will take many years and will cost billions of dollars. Because of the higher skill and training requirements for survey technicians, wages will increase and more individuals will seek to enter the profession. Continuing education will be mandatory if you intend to stay ahead of the competition. As stated earlier in this book, membership in your state Land Surveyors Association and the American Congress on Surveying and Mapping are excellent ways to keep current with new ideas and techniques.

You have taken your first step in a new profession by completing this course. As you gain experience, you will find many opportunities for advancement. Through continuing your education and doing your best work, you will be in an excellent position to take advantage of them. Good luck!

## Review Questions

1. A. What is a TSSS?
   B. What are the component parts of a TSSS?
2. What is raw field data?
3. A. With so many good calculators available, why is it necessary to know how to do the calculations by hand?
   B. Why not just use the calculator program?
4. Define GIS.
5. Define LIS.
6. A. In your own words, state the benefits of a GIS/LIS to your area of the country.
   B. What are some of the problems you see in doing a GIS/LIS in your area?

# Surveyor's Hand Signals

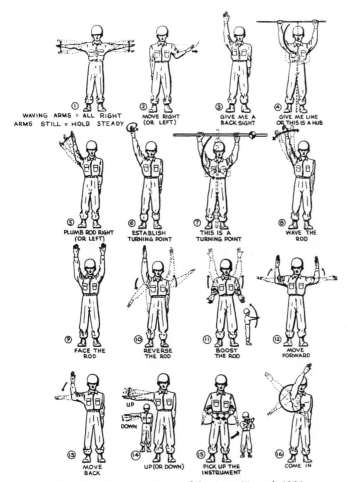

① WAVING ARMS = ALL RIGHT
ARMS STILL = HOLD STEADY

② MOVE RIGHT
(OR LEFT)

③ GIVE ME A
BACK SIGHT

④ GIVE ME LINE
OR THIS IS A HUB

⑤ PLUMB ROD RIGHT
(OR LEFT)

⑥ ESTABLISH
TURNING POINT

⑦ THIS IS A
TURNING POINT

⑧ WAVE THE
ROD

⑨ FACE THE
ROD

⑩ REVERSE
THE ROD

⑪ BOOST
THE ROD

⑫ MOVE
FORWARD

⑬ MOVE
BACK

⑭ UP (OR DOWN)

⑮ PICK UP THE
INSTRUMENT

⑯ COME IN

*Courtesy of the Department of the Army Manual, 1964.*

# Survey Control Offices and Bureau of Land Management Offices

Information on the field notes, plats, maps, and other papers relating to the Surveys of the Public Lands of the United States can be obtained for each state from the following:

Alabama-Secretary of State,
Montgomery, AL 36104

Arkansas-Department of State Lands,
State Capitol,
Little Rock, AR 72201

Florida-Board of Trustees of the Internal Improvement Trust Fund,
Elliott Bldg.,
Tallahassee, FL 32304

Illinois-Illinois State Archives,
Secretary of State,
Springfield, IL 62706

Indiana-Archivist, Indiana State Library,
140 North Senate Avenue,
Indianapolis, IN 46204

Iowa-Secretary of State,
Des Moines, IA 50319

Kansas-Auditor of State and
Register of State Lands,
Topeka, KS 66612

Louisiana-Register, State Land Office,
Baton Rouge, LA 70804

Michigan-Department of Treasury,
Bureau of Local Government Services,
Treasury Building, Lansing, MI 48922

Minnesota-Department of Conservation,
Division of Lands and Forestry,
Centennial Office Building,
Saint Paul, MN 55101

Mississippi-State Land Commissioner,
P.O. Box 39,
Jackson, MS 39205

Missouri-State Land Survey Authority,
P. O. Box 1158,
Rolla, MO 65401

Nebraska-State Surveyor,
State Capitol Building, P.O. Box 4663,
Lincoln, NE 68509

North Dakota-State Water Conservation Commission,
State Office Building,
Bismarck, ND 58501

Ohio-Auditor of State,
Columbus, OH 43215

South Dakota-Commissioner of School
and Public Lands, State Capitol,
Pierre, SD 57501

Wisconsin-Department of Natural Resources,
Box 450,
Madison WI 53701

State offices of the Bureau of Land Management:
Alaska, Anchorage
Arizona, Phoenix
California, Sacramento
Colorado, Denver
Idaho, Boise
Maryland, Silver Spring
Montana, Billings
Nevada, Reno
New Mexico, Santa Fe
Oregon, Portland
Utah, Salt Lake City
Washington, Portland, Oregon
Wyoming, Cheyenne

# Common Survey Abbreviations

**1/16**  sixteenth section
**1/4**  quarter section
**ACSM**  American Congress on Surveying and Mapping
**ALTA**  American Land Title Association
**AM**  amended monument
**AMC**  amended meander corner
**AP**  angle point
**ASPRS**  American Society of Photogrammetry and Remote Sensing
**BC**  begin curve
**BCR**  begin curb return
**BLM**  Bureau of Land Management
**BM**  bench mark
**BO**  bearing object
**BOB**  basis of bearing
**BT**  bearing tree
**BVC**  begin vertical curve
**C**  center
**CADD**  Computer Aided Drafting and Design
**CC**  closing corner
**CL ( ₵ )**  center line
**CO-GO, COGO**  Coordinate Geometry
**D**  degree of curve
**DMD**  double meridian distance
**DUT1**  UTC correction for your time zone (UT1)
**dbh**  diameter breast high (for trees), 4.5 ft. from grade
**E**  east
**E**  external distance
**EC**  electronic control

**EC**  end curve
**ECR**  end curb return
**EDM**  electronic distance measuring
**EP**  edge of paving
**EVC**  end vertical curve
**FGCC**  Federal Geodetic Control Committee
**GIS**  Geographic Information System
**GIS/LIS**  Geographic Information System/Land Information System
**GLO**  Government Land Office
**GPS**  Global Positioning System
**HI**  height of instrument
**I**  intersecting angle between tangents of a curve
**L**  length of curve
**LC**  long chord
**LCD**  liquid crystal display
**LED**  light emitting diode
**LIS**  Land Information System
**LM**  local monument
**LS**  registered land surveyor
**M**  middle distance, midordinate for a curve
**M**  mile
**MC**  meander corner
**MPVC**  midpoint vertical curve
**N**  north
**NAD 83**  The World Geodetic System of 1983
**NAVD 88**  North American Vertical Datum 1988
**NAVSTAR**  NAVigation Satellite Time and Ranging GPS System
**NE**  northeast
**NGS**  national geodetic survey

279

**NOAA** National Oceanic and Atmospheric Administration
**NOS** national ocean survey
**NW** northwest
**PC** point of curvature (beginning of curve)
**PE** professional engineer
**PI** point of intersection
**PL** ( ℞ ) property line
**POB** point of beginning
**POC** point on curve
**POL** point on line
**POVC** point on vertical curve
**PT** point of tangency (end of curve)
**R** radius (as in a horizontal or spiral curve)
**R** range
**RM** reference monument
**ROS** record of survey
**RP** reference point
**S** section

**SC** standard corner
**SE** southeast
**SMC** standard meander corner
**SW** southwest
**T** tangent distance
**T** township
**TBM** temporary bench mark
**TO** ( $\bar{\Phi}$ ) throw over (plunged) ties
**TR** tract
**UTC** Coordinated Universal Time
**UT1** zonal correction to UTC (or DUT1)
**VBM** vertical angle bench mark
**VCP** vitrefied clay pipe
**VPI** vertical point intersection
**W** west
**WC** witness corner
**WP** witness point
**Δ** delta

# Glossary

ACCRETION. The gradual and imperceptible adding to the shore.

ACCURACY. The degree of conformity with a standard or accepted value; how close you are to the true measurement.

ACQUIESCENCE. A concurrence by adjoining owners bearing on the location of their common boundary.

ACT OF NATURE. Not caused by mankind.

ADVERSE POSSESSION. A method of overcoming written title, acquiring title by possession for a statutory period.

AGONIC. A line on the earth's surface joining points that have no declination; the compass needle points north or south.

ALIDADE/UPPER MOTION. (a) The upper unit of a transit. (b) An instrument with a telescope mounted to a straight edge, used with a plane table for taking topography.

ALIQUOT. A part of a distance that divides the distance evenly, without a remainder.

ALLUVION. The material deposited by water.

ALTIMETER. (Figure 4-4) An instrument used for measuring elevation and height.

AMBIGUITY. An uncertainty: *Latent* does not appear but comes from evidence. *Patent* appears on the instrument and causes it to be defective.

ANGLE OFFSET OR OFFSET. (Figure 3-14) A way to go around an object (*see* OFFSET).

ANALYTICAL PHOTOTRIANGULATION. A precise method of finding distance from air photos.

APPURTENANCE. That which belongs or is attached to something else, an accessory.

ARPENT. A measure of an area approximately 0.85 acres: when used as a linear description, an arpent refers to a length along a side that will give 1 square arpent of land. The values differ between French and English grantors. Local experts should be used to determine the length in your area.

"AS BUILT." A survey made to show any difference between the designed and finished project.

ASPIRATION. Evaporation of water into the atmosphere.

ASTROLABE. (Figure 2-25) An instrument for measuring altitudes of celestial objects, from the Greek words "to take a star."

AUTOMATIC/SELF-LEVELING LEVEL. (Figure 2-24) A level that keeps the line of sight level through use of a prism pendulum.

AZIMUTH. (Figure 6-4) An angle rotated clockwise 360° from either north or south meridian.

BACKSIGHT. (Figure 4-9) (a) A point used to determine elevation or angular orientation of the instrument. (b) In leveling, the plus sight if elevation is lower than line of sight.

BALANCE (IN TURNS/SHOTS). (Figure 4-12) (a) Distributing corrections through a traverse to eliminate errors of closure. (b) Equal distances for backsights and foresights in leveling.

BEARING. (Figure 6-5) The angle measured from a north or south meridian in either an east or west direction, maximum of 90°.

BEARING ROTATION. (Figure 6-7) Always begins north or south and rotates either east or west and cannot exceed an angle of 90°.

BEARING TREE. (Figure 12-9) Marked by a blaze with BT, section, township, range in vertical order reading downward; location, size, and species are recorded.

BENCH MARK. A permanent object, natural or artificial, marked with an elevation point.

BLUNDER. A mistake; reversing numerals, failing to record a measured distance, or incorrectly reading a horizontal circle—new measurements are usually required.

BOOTING THE ROD. (Figure 8-4) Raising the rod a carefully measured amount to extend the line of sight.

BORROW PIT. (Figure 11-12) The pit from which fill material is removed.

BREAKING TAPE/CHAIN. (Figure 3-9) Short distances measured and accumulated to total a full tape length when a standard 100-ft. tape cannot be held horizontally without plumbing from above shoulder level.

BUBBLE/FISH EYE. A circular bubble used to determine plumb.

CADASTRAL SURVEY. Relates to land boundaries and subdivisions, usually used to designate the surveys of the public lands of the United States.

CARDINAL DIRECTION. The direction on the surface of the earth—north, south, east or west; sometimes indicates all of them.

CATCH POINT/DAYLIGHT POINT. The point where the cut or fill slope intercepts the natural ground.

"CHAIN!" Used as a command it means *STOP NOW!*

CHAINING/TAPING. Measuring distance on the ground with a tape or chain; formerly, the terms were used synonymously but "taping" is now preferred, except in public land surveys.

CHORD

LONG. (a) A straight line that extends from the point of curvature to the point of tangency. (b) In route surveying, indicates a straight line between two points on a curve, regardless of distance between them. (c) In a description of a circular land boundary, the length and bearing of the long chord is an important factor.

SHORT. A straight line connecting two points on a curve.

COMPASS. Used to determine azimuth or direction relative to the meridian.

MAGNETIC. (Figure 6-1) Uses a magnetic needle to indicate magnetic north.

SOLAR (BURT'S IMPROVED). (Figure 1-12) Permits instantaneous mechanical solution of the astronomic triangle (sun-zenith-pole), replaced by transit.

CONE OF VISIBILITY. The area visible to a camera lens during aerial photography, it increases with the altitude of the flight.

CONTOUR. (Figures 8-14, 15) (a) An imaginary line on the ground, all points of which are at the same elevation above a specific surface. The datum surface most generally used in the United States is mean sea level. (b) The lines showing changes in elevation on a topography map.

CONTROL SURVEY. Provides horizontal or vertical position data for the support of subordinate surveys or mapping.

CONVEYANCE. A document that passes an interest in real property from one person to another as in a deed, mortgage, or lease.

CORNER. A point on a boundary at which two or more boundary lines meet.

EXISTENT. Position can be determined by original monument or accessories.

LOST. Position cannot be determined beyond reasonable doubt or methods used for obliterated corners.

OBLITERATED. All trace of original monument and accessories are lost, but position can be determined by local authorities or records.

CORNER ACCESSORIES. Can be timber scribed trees, cliffs, boulders, stone mounds, pits, or memorials in the ground that can be clearly marked.

CROSS SECTION. (Figure 8-18) A horizontal grid system used to determine contours, quantities of earthwork, etc., by means of elevations of the grid points.

CULVERT. A ditch passing under a structure; pipe or box culverts, made of galvanized iron or reinforced concrete.

CUL-DE-SAC. (Figure 14-13) A dead-end street that is wide enough at its end to permit a U-turn by a fire truck.

CURVATURE. (Figure 4-2) Curvature of the earth is roughly 0.02 ft. in 1000 ft. and keeps a level line from being flat.

CURVES, HORIZONTAL. (Figure 10-5)

COMPOUND. (Figure 10-2) Two simple curves of different radii joined together and curving in the same direction.

REVERSE. (Figure 10-3) Two simple curves joined together but curving in opposite directions.

SIMPLE. (Figure 10-1) An arc of a circle, the radius determines the sharpness or flatness of the curve.

SPIRAL. (Figure 10-4) Has a constantly changing radius, provides a gentle transition from the tangent to a simple curve or joins two simple curves in a compound curve.

CURVES, VERTICAL. (Figures 10-10, 11, 13)

HUMP/CREST/SUMMIT. (Figure 10-9) A curve that begins upward and ends downward: $G^1$ is positive, $G^2$ is negative.

SAG/DEPRESSION. (Figure 10-9) A curve that begins downward and ends upward: $G^1$ is negative, $G^2$ is positive.

DATA COLLECTOR. (Figures 2-36, 15-2) Records the readings from an electronic theodolite as they are taken and can download directly to a computer.

DAYLIGHT POINT/CATCH POINT. (Figure 11-13) (see CATCH POINT.)

DECLINATION. (a) The angle between true north (astronomic) and magnetic north, (b) The declination of the sun or a star is the angle above or below the celestial equator.

DEED. An instrument in writing that, when properly executed and delivered, conveys an estate in real property or in interest in that property.

DEFLECTION ANGLE. (Figure 7-3) The angle measured by sighting at a point, then transiting the scope through a vertical angle of 180° and turning to a new point, right or left of the projected line that is ahead of the instrument point.

DIFFERENTIAL LEVELING. Measuring the difference of elevation between any two points by spirit leveling.

DOUBLE MERIDIAN DISTANCE (DMD). (Figure 7-13) A method of finding area; uses latitudes and departures of the adjusted traverse to calculate the area by using the area of a trapezoid (see Chapter 7).

DOUBLE PROPORTIONATE MEASUREMENT. (Figure 12-5) A method used to restore lost corners; the distances between four found corners, two in each cardinal direction from the lost corner, are used to find a lost corner common to four townships or the interior corner of four sections.

DOUBLE RODDING. (Figure 4-16) Used for long level runs, using a plus rod and a minus rod, records turns as high and low shots.

DTM (DIGITAL TERRAIN MODELING). Used by computers to construct topographic maps.

DUMPY/ENGINEER'S LEVEL. (Figure 2-23) Similar to a wye level except the telescope is fastened to the base.

EASEMENT. Allows one person partial use for a specific purpose, but not possession, of the land belonging to another person.

EDM. (Figure 2-10 and others) Electronic distance measurement; a precision surveying instrument.

ELLIPSOID. The mathematical surface approximately the shape of the Earth.

EMINENT DOMAIN. The right of cities, counties, public utilities, highway departments, parks, school districts, railroads, pipelines, and so forth to take private property for public use; just compensation must be given.

ENGINEER'S TAPE. (Figure 2-8) Also known as a chain, a band of metal or alloy used for measuring the lengths of base lines, traverse lines, controlling triangulation, etc.

EROSION. The slow and imperceptible wearing away of land by water action.

ERRORS

ACCIDENTAL. Random, known as plus/minus errors, as in rounding off numbers; tend to cancel themselves but can be large or small, cannot be measured or removed.

BLUNDER. A mistake that cannot be anticipated but must be corrected.

SYSTEMATIC. Cumulative; caused by known deviations such as temperature, curvature, refraction, or index error in vertical reading, removed by adjustment or a formula.

EVIDENCE. Testimony, writings, material objects, or other things presented to prove the existence or nonexistence of a fact.

PAROLE. Given by witnesses that saw the location of real evidence when it existed, should be recorded in the field book and witnessed.

REAL. Monuments, fences, roads, signs of continued use, etc.

WRITTEN. Titles, maps, public records documents pertaining to boundaries.

FISH EYE/BUBBLE. The circular bubble used to determine plumb.

FLY POINT. (Figure 7-12) A point outside a closed

traverse that is connected to the traverse by distance and angle, an open traverse.

FORESIGHT. (Figure 4-9) The second sight; after the back sight; the minus shot in leveling if elevation is below line of sight.

FRISCO ROD. (*See* ROD, FRISCO.)

GEOID. The imaginary sea level surface of the Earth perpendicular to the plumb line at all points.

GAMMON REEL. A device used to control the plumb bob string.

GLOBAL POSITIONING SYSTEM (GPS). (Figure 3-16) A satellite surveying system; three-dimensional measurement based on observation of radio signals of the NAVSTAR GPS system.

GRADIENT. Grade of slope or rise over run.

GRID TOPO. A topo worked on an area usually divided into a 100-ft. grid, using lath and flagging; best in an area that has little relief and is fairly free of trees.

GROMA. An ancient surveying instrument used to lay out right angles by sighting through a pair of plumb lines.

GUNTER'S CHAIN. (Figure 2-7) Used to measure distance, composed of 100 links, each link being 7.92-in. long, for a total of 66 ft. in length.

HAND LEVEL/P-GUN. (Figure 2-4) A hand-held level whose name probably comes from the "P" lines in a preliminary survey.

HEIGHT OF INSTRUMENT (HI). The distance between the station or hub and the center of the instrument barrel. In leveling, refers to the elevation of the instrument.

HEMISPHERE. One half a circle.

HORIZONTAL ANGLE. (Figure 7-3) Three types of angles on a horizontal plane; right, left, and deflection.

HORIZONTAL LINE. (Figure 4-1) Forms a 90° angle with a vertical line.

HUB. A square stake, driven flush with the ground, with a tack marking a survey point.

HUMP/CREST/SUMMIT. (*See* CURVE, VERTICAL.)

INVERT. The lowest part of a tunnel or the flow line pipe in a sanitary sewer.

ISOGONIC CHART. (Figure 6-3) Shows the annual changes in magnetic declination of the magnetic north pole.

ISOGONIC LINE. A line joining points of equal declination on the earth's surface.

KEEL/KIEL/LUMBER CRAYON. Used for marking lath or hubs, sold in hardware stores as lumber crayon.

KINEMATIC (IN MOTION) GPS SURVEYING. Uses a mobile receiver that is calibrated to a known position. (*See* article in Chapter 3 from Trimble Navigation.)

LAPLACE CORRECTION. Applied to an azimuth to compensate for deflection of the plumb. Measurements are taken at LaPlace Stations and corrections are published by the government.

LATH. A 1/4 in. by 2 in. by 2 ft. or 4 ft. wood slat used as a marker.

LEAST SQUARES ADJUSTMENT. Uses the minimum sum of the weighted residuals squared to arrive at the most probable measurement in 2-D or 3-D traverse adjustments.

LEVEL LINE. (Figure 4-2) Curves to follow sea level, do not confuse with a horizontal line.

LEVEL LOOP. A level run starting from a bench mark and closing on the same bench mark.

LEVELS

AUTOMATIC. (Figure 2-24) It keeps the line of sight level through use of a prism pendulum.

DUMPY. (Figure 2-23) Similar to the WYE level except the telescope is permanently fastened to the base.

ENGINEER'S. Same as a DUMPY level.

SELF-LEVELING. Same as an AUTOMATIC level.

SPIRIT. Circa 1666, had curved glass tube for smooth bubble movement.

TILTING/SPIRIT BUBBLE. Used for precise work, both ends of the bubble are visible in the field of view, final adjustment by micrometer knob.

WYE. (Figure 2-22) 1800s spirit level, bubble mounted on tripod, telescope above it on wyes could be lifted out and reversed.

LENKER ROD. (Figure 2-19) (*See* ROD, LENKER.)

LITTORAL. Pertains to ownership on a lake or a seashore.

LOCAL ANOMALIES. Variations that cause a deflection from the vertical, can be compensated for by use of the LaPlace correction, magnetic anomalies-local attraction.

LOCAL COORDINATE SYSTEM. The coordinates assigned by the surveyor when no coordinates are given.

MAGNETIC COMPASS. (Figure 6-1) (*See* COMPASS, MAGNETIC.)

MAGNETIC NORTH OR SOUTH. The direction of a north-seeking end of a magnetic compass needle not subject to transient or local disturbance.

MAPS

CADASTRAL. Shows land boundaries and subdivisions, usually used to designate surveys of public lands in the United States.

FINAL. Shows changes required by planning commission, used to stake out improvements and lot corners for construction once it is approved.

GRADING. Shows difference in elevation between existing elevation and final grade.

HYDROGRAPHIC. Shows the land beneath the surface of the water.

ISOGONIC. (Figure 6-3) Shows annual changes in declination.

PHOTOGRAMMETRIC. Made from aerial photographs, may be topographical or orthometric in which all points are projected perpendicular to the earth's surface providing an undistorted view.

PLANIMETRIC. Shows the horizontal positions of features as seen from above.

PROFILE. Shows vertical differences in elevation as seen from the side view.

TENTATIVE. Shows proposed layouts of lots and other improvements submitted for planning commission approval.

THEMATIC. Shows a special theme or feature, like population, soil type, spread of disease, etc.

TOPOGRAPHIC. (Figures 8-14, 15) Shows the land in three dimensions, differences in elevations shown by contour lines; (Figure 8-9) Computers use DTM programs to draw contours.

MAP/PLAN/PLAT. A view of a plane surface, at an established scale, of the physical features of a part of the whole of the earth's surface, by use of signs and symbols and with a method of orientation indicated.

MEANDER. The turning or winding of a stream.

MEANDER LINE. The survey line set to separate the public lands from the body of water.

MERIDIANS, PRINCIPAL. (Figure 12-8) A line extending along the astronomical meridian, along which township and north-south lines are established.

METES AND BOUNDS DESCRIPTION. Describes a parcel of land by references to course and distances around a tract, to natural or record monuments, or by reference to adjacent land owners.

MINUS ROD. (Figure 4-10) A foresight, any number can be taken in a line of levels, assuming the elevation is lower than the line of sight.

MIRROR/PRISM/RETROPRISM. (Figures 2-12, 13, 14) Designed to reflect a beam of light back parallel to its entrance angle.

MONUMENT. (Figure 12-4) A physical structure that marks the location of a corner or other survey point. It may be natural or man-made.

NADIR. (Figure 4-1) Also called a plumb point, the opposite of zenith; the point on the ground or at sea level vertically beneath the perspective center of the camera lens or photo image of that point.

NAVIGABLE. (The courts are undecided on this but the best definition seems to be) Water capable of being traversed by a fully laden cargo canoe at the point of lowest low water.

NORTHING/EASTING. A linear distance; the coordinates of a point.

OBVERT. (Figure 4-11) The elevation of the ceiling or top of a tunnel or sewer pipe.

OFFSET OR ANGLE OFFSET. (Figure 3-14) A way around an object.

TANGENT METHOD. (Figure 12-6) Best for open country where brushing is not necessary.

SECANT METHOD. (Figure 12-7) Best for timbered country; does not require long offsets.

OPTICAL PLUMMET. A device for plumbing an instrument over a point.

ORTHOPHOTO. An aerial photograph that has been corrected for tilt and relief.

P-GUN/HAND LEVEL. (See HAND LEVEL.)

PARALLAX. The apparent movement of the eyepiece cross hairs across an object, caused by improper focus of the eyepiece on the cross hairs.

PEANUT MIRROR/1-INCH PRISM. (Figure 3-13) A reflecting device that is sometimes mounted on a base that can be screwed into an extended pole, sometimes having a hook at the base for suspending a plumb bob.

PEG NOTES. Unofficial notes written on any paper other than the field book.

PEG TEST FOR LEVEL. (Figure 4-19) A method of adjusting a level to reduce error in horizontal line of sight.

PERCOLATION. The natural seeping of water into the soil.

PERSONAL HI. The distance from the ground to the observer's eye.

PERSPECTIVE. All points converge at a single point like the lens of a camera. Distances are distorted. An example is an aerial photograph.

PHILLY ROD. (See ROD, PHILLY.)

PHOTOGRAMMETRY. (Figure 8-8) "The art, science and technology of surveying and measuring by photographic and other energy emitting process" (ASPRS).

PLAT. A drawing of the area included in a survey.

PLOTTING. (Figures 8-10, 11) Preparing a plat from field notes.

PLUMB BOB. (Figure 2-3) A conical brass weight suspended by a string, used to project a point vertically.

PLUS ROD. A rod read from the bench mark or backsight at the beginning of a level run.

POINT ON LINE (POL). Point on the line of survey.

POLARIS. (Figure 6-13) The North Star.

PRECISION. Consistency, uniformity; to measure something repeatedly and have all the measurements fall within the area of probability for the true measurement; a combination of instrument quality and operator ability.

PREMARKS. (Figure 8-7) Large white crosses painted on paving or marked with fabric on the ground to guide the pilot in flight; for use by a photogrammetrist in aerial mapping.

PRESCRIPTIVE RIGHT. The right to an easement across another person's property; similar to adverse possession.

PRISMOIDAL FORMULA. Used to compute volume.

PRISM/MIRROR. (*See* MIRROR/PRISM.)

PROFILE. A side view of a project.

PRORATE. To divide or distribute proportionally.

PYTHAGOREAN THEOREM. The sum of the squares of the two sides of a right triangle is equal to the square of the hypotenuse; $a^2 + b^2 = c^2$.

QUADRANT. One fourth of a circle; starting at north and going clockwise, they are: 1st, northeast; 2nd, southeast; 3rd, southwest and 4th, northwest.

RANGE LINES. Meridional lines for public land surveys; hydrographic mapping also uses meridional lines.

RANGE POLE. (Figure 2-15) Red and white wood or plastic-coated metal pole, striped at one foot intervals; provides alignment sights for taping and transit.

RECIPROCAL LEVELING. (Figure 4-17) A procedure of measuring vertical angles or making rod readings from two instruments to compensate for instrument misadjustment or turns forced out of balance by intervening obstacles.

RECTANGULAR COORDINATE SYSTEM. Coordinates on any system in which the axes of reference intersect at right angles.

RECTANGULAR SYSTEM OF SURVEY. (Figure 12-1) A survey of an area divided by base lines intercepted at regular intervals by meridional lines. Lines usually run north and south for meridians, east and west for bases.

REFERENCE POINTS
    SWING TIES. (Figure 9-1) Can be set with only a tape.
    THROW OVER TIES. (Figure 9-2) Must be set with an instrument and a tape.

REFRACTION. (Figure 4-2) The bend of light waves as they pass through the earth's atmosphere.

RELICTION. Exposing of the land by receding water as in a lake drying up.

REPOSE. Stability of a slope.

RETROPRISM. (*See* MIRROR/PRISM.)

REVULSION/AVULSION. The sudden tearing away of the land by water.

RIGHT ANGLE PRISM. (Figure 2-17) Used to find an approximate perpendicular to your line of sight.

RIPARIAN. Ownership relating to a boundary by a river or stream.

ROD. A measurement of length, also known as a perch or pole when used for square measurement; the official length of a rod in the United States is 16.5 ft. It is not the same thing as a level rod.

RODS/LEVELING
    FRISCO/SAN FRANCISCO. A three-piece rod.
    INVAR. Matched pairs, made of low expansion metal, very expensive.
    LENKER. (Figure 2-19) A direct reading rod.
    PHILLY/PHILADELPHIA. A two-piece rod.
    STADIA. (Figure 2-21) A folding rod, graduated in feet, tenths, and half tenths, easier to read at long distances.
    TELESCOPING. (Figure 2-20) Available in lengths of 25 or more ft.

SAG/DEPRESSION. (*See* CURVE, VERTICAL.)

SENIOR RIGHTS/JUNIOR RIGHTS. (Figures 13-4, 5, 6) First buyer (senior) gets all land deeded by grantor, second buyer (junior) gets remainder.

SLOPE CHAINING/TAPING. (Figure 3-10) Tape is held, by design, out of horizontal; measurement is made by using trigonometry and the difference in elevation.

SOLAR COMPASS (BURT'S IMPROVED). (*See* COMPASS, SOLAR.)

SPANDREL POINTS. (Figure 11-11) The intersection of the flow lines at a street intersection.

SPRING BALANCE. (Figure 3-6) Used to provide proper tension to a suspended tape, about 20 lbs. per 100 ft.

STADIA. A method of reading distance using two hairs in the field of view of a telescope that are set at a proportional distance to the observed distance on a rod. Usually a 1 : 100 proportion or 1 ft. on the rod equals 100 ft. between the instrument and the rod.

STADIA BOARD/ROD. (*See* ROD, STADIA.)

STADIA SHOT. (Figure 4-20) Reading distance by using the upper and lower cross hairs in the instrument.

STATE PLANE COORDINATES. The plane-rectangular coordinate system established by the U.S. Coast and Geodetic Survey – at least one for each state.

STRIP TOPO. An elongated topography map used for sewer or street design.

SUBDIVISION DESCRIPTION. (Figure 12-3) The division of land into lots, tracts, parcels sites, or divisions for purpose of sale, lease, or transfer of ownership. Check local regulations.

SUMMIT/UPWARD VERTICAL CURVE. (*See* HUMP/ CREST/SUMMIT.)

TALWEG/THALWEG. The center of the deepest or lowest part of the main channel of a river or stream.

TANGENT. (1) A straight line that touches a curve at one point only and does not intersect it; (2) In route alignment: (a) part of the alignment from one PI to the next PI; (b) the distance from the PT of one curve to the PC of the next.

TAPE/CHAIN. A ribbon of steel, invar, cloth or other suitable material, graduated into units of length, used to measure distances. It replaced the Gunter's chain in surveying in 1900.

TECH/TECHNICIAN. A subprofessional expert in the field of surveying technology.

TEMPERATURE CORRECTION. (Figure 3-8) Tables that show corrections for expansion or contraction of a steel tape due to temperature; approximately .01 per 100 ft. per 15° of Fahrenheit.

THEODOLITE. A precision surveying instrument used to measure horizontal and vertical angles. There are two types of theodolites:

DIRECTIONAL. The horizontal circle remains fixed during a series of observations, used for first and second order triangulations.

REPEATING. The horizontal measurement can be accumulated on the circle. A mean angle can be calculated by dividing the accumulated angle by the number of repetitions. Used for most regular, day-to-day surveying.

THROW OVER/TRANSIT. (*See* TRANSIT.)

TOE. Bottom point on a slope.

TOPOGRAPHY. The features of the surface of the earth: Hypsography – the relief, and Hydrography – the water, culture, and man-made features.

TOTAL STATION. An instrument that combines a theodolite with an EDM and a calculator.

TOWNSHIP. (1) A quadrangle of land 6 miles on a side, (2) A unit of local government.

TOWNSHIP LINES. (Figure 12-2) The exterior of a township, a 6-mile square.

TRACE CONTOUR. To follow a contour line with a level rod.

TRANSIT/THROW OVER, PLUNGE THE SCOPE. (1) An instrument used to measure horizontal and vertical angle; called a transit because of the ability to rotate the telescope about a horizontal axis (throw over). Angles are read from a vernier mounted next to the horizontal and vertical circles; (2) The movement of a celestial body across a defined line on the celestial sphere, such as a meridian.

TRAVERSE. A series of lines, of measured length, connected by measured angles.

TRAVERSE RULES. Methods of adjusting a traverse to minimize the effects of errors.

COMPASS (OR BOWDICH). Assumes the error is caused equally by angle and distance; most commonly used balancing method, also in computer programs.

CRANDALL. Assumes error is greater in measurement of distance or angles and uses a modified least squares adjustment for better results than the transit rule; common to computer adjustment programs.

LEAST SQUARES. The sum of the squares of all the corrections or residuals derived for the observed data is made a minimum, based on the theory of probability.

OPEN. Begins from a station of known or adopted position, but does not end on such, also termed an open-end traverse.

TRANSIT. Assumes error is greater in measurement of distance than angles.

TRAVERSE TYPES

CLOSED. (Figure 7-2) Begins and ends on a known position.

CONNECTING. (Figure 7-8) Begins on a known position/point, goes through a point or points with uncertain position(s) and ends on a point whose position is known.

TRIBRACH. The unit used to fasten the theodolite to the tripod. The theodolite may be removed from the tribrach by means of a locking cam, controlled by a knob on the side of the tribrach.

TRIGONOMETRIC LEVELING. Use of the sine function to find the difference in elevation.

TRUE NORTH OR SOUTH. (1) An astronomic meridian;

(2) a geodetic meridian; (3) the direction of north from magnetic north corrected for declination, (4) the meridional direction assumed in a survey description; (5) the cardinal directions run in public land surveys.

TURNING POINT. (Figure 4-8) A specific point that allows only one point to contact the footplate of the rod.

UNWRITTEN CONVEYANCE. Transfer of property by oral or implied agreement.

UPPER MOTION/ALIDADE. (See ALIDADE.)

VARA. Old Spanish unit of measure used in southwestern United States and Mexico; length varies with the state using it from 32.8 to 39.6 varas per 100 ft.

VARIATION/DECLINATION. The angle, east or west, of the true pole to the magnetic pole caused by the offset of the magnetic pole from the true pole.

VERNIER. (Figure 2-31) An auxiliary scale sliding against and used in reading the primary scale.

WIGGLE IN/BUCK IN. (Figure 5-11) An intermediate point set when a point-to-point view is not possible.

WIND ANGLE/WING DING. (See WING DING.)

WINDING UP AN ANGLE. (Figure 5-5) Repeating the angle.

WING DING/WIND ANGLE. (Figure 5-1) A 90° angle formed by standing aligned along a line and bringing both arms together to form an angle perpendicular to the line.

WITNESS POINT/WITNESS CORNER. (See EVIDENCE, PAROLE.)

WYE LEVEL. (See LEVEL, WYE.)

ZENITH. (Figure 5-8) The direction straight overhead, or vertically above the instrument or survey point.

# Bibliography

## Chapter 1

Brown, C. M., and Eldridge, W. H. *Evidence and Procedures for Boundary Location.* New York: John Wiley & Sons, 1967.

de Camp, L. Sprague. *The Ancient Engineers.* New York: Ballentine Books, 1984.

*Encyclopaedia Britannica.* Chicago, IL: Encyclopaedia Britannica, Inc., 1975.

Keily, Edmond R., PhD. *Surveying Instruments: Their History.* Columbus, OH: Carben Surveying Reprints, 1979.

*Surveying and Mapping, Journals and Bulletins.* Bethesda, MD: American Congress on Surveying and Mapping, 1980–1990.

Viola, H.J. *Exploring the West.* Washington, DC: Smithsonian Press, 1987.

White, C.A. *A History of the Rectangular Survey System.* U.S. Department of Interior Bureau of Land Management, 1980.

## Chapter 2

Keily, Edmond R., PhD. *Surveying Instruments: Their History.* Columbus, OH: Carben Surveying Reprints, 1979.

*Surveying with GPS.* Sunnyvale, CA: Trimble Navigation, 1989.

U.S. Department of the Army Technical Manual TM5-232. *Elements of Surveying,* August, 1964.

## Chapter 3

Mooyman, Ken, and Quiron, Cheryl A. *High Production Kinematic GPS Surveying.* Workshop pamphlet, 1990.

Reilly, James P., PhD. *P.O.L.* Portland, OR: Portland Precision Instruments & Repair Co., 1989.

Trimble Navigation, Ltd. Pamphlets. Sunnyvale, CA: 1990.

## Chapter 4

Federal Geodetic Control Committee Manual, U.S. Department of Commerce. *Standards and Specifications for Geodetic Control Networks.* September, 1984.

## Chapter 5

U.S. Department of the Army Technical Manual TM5-232. *Elements of Surveying.* August, l964.

## Chapter 6

Bouchart, Harry, and Moffitt, Francis H. *Surveying.* 5th ed. Scranton, PA: International Textbook Co., 1970.

Cothern R. *Solar Observations and Calculations.* Anchorage, AK: University of Alaska, 1984.

Kiely, Edmond R., PhD. *Surveying Instruments:*

*Their History.* Columbus, OH: Carben Surveying Reprints, 1979.

Signani, Larry. *Magnetic Surveys and Instruments.* Seattle, WA: U.S. Corps of Engineers, 1988.

## Chapter 7

Bouchard, Harry, and Moffitt, Francis H. *Surveying.* 5th ed. Scranton, PA: International Textbook Co., 1970.

Buckner, R. B. *Surveying Measurements and Their Analysis.* Rancho Cordova, CA: Landmark Enterprises, 1986.

Star*Net. Software bulletins. Oakland, CA: Starplus Software, 1990.

## Chapter 8

*American Land Title Association seminar pamphlet.* San Francisco, CA: 1974.

Keily, Edmond R., PhD. *Surveying Instruments: Their History.* Columbus, OH: Carben Surveying Reprints, 1979.

PacSoft software. *Informational pamphlets.* Kirkland, WA: Pacsoft, 1990.

*Pre-Marking for Aerial Photography* bulletin. Bellevue, WA: Nies Mapping Group, Inc., 1990.

U.S. Department of the Interior Bureau of Land Management. *Standard Field Tables.* 1956.

## Chapter 9

Buckner, R. B. *Surveying Measurements and Their Analysis.* Rancho Cordova, CA: Landmark Enterprises, 1986.

Corvalis Microtechnology Inc. Data brochure. Corvalis, OR: 1990.

Federal Geodetic Control Committee Manual, U.S. Department of Commerce. *Standards and Specifications for Geodetic Control Networks.* September, 1984.

Hewlett-Packard informational phamplets. Palo Alto, CA: 1990.

Mikhail, Edward M., PhD., and Gracie, Gordon, PhD. *Analysis and Adjustment of Survey Measurement.* New York: Van Nostrand Reinhold, 1981.

U.S. Department of Commerce National Oceanic and Atmospheric Administration. *State Plane Coordinate System of 1983.* January, 1989.

*Surveying and Mapping* Bulletins and Manuals. Bethesda, MD: American Congress on Surveying and Mapping, 1980–1990.

## Chapter 10

Hewlett-Packard informational pamphlets. Palo Alto, CA: Hewlett-Packard, 1990.

Meyer, Carl F. *Route Surveying and Design.* 4th ed. Scranton, PA: International Textbook Co., 1971.

U.S. Department of the Army Technical Manual TM5-232, *Elements of Surveying.* 1964.

Washington State Department of Transportation. *Field Tables for Engineers.* 1984.

## Chapter 11

PacSoft software. Informational pamphlets. Kirkland, WA: PacSoft, 1990.

## Chapter 12

Brown, C. M. *Boundary Control and Legal Principles.* New York: John Wiley & Sons, 1969.

Brown, C. M., and Eldridge, W. H. *Evidence and Procedures for Boundary Location.* New York: John Wiley & Sons, 1967.

Grimes, John S. *Clark on Surveying and Boundaries.* Charlottesville, VA: The Michie Co., 1976.

U.S. Department of the Interior. *Manual of Instructions for the Survey of Public Lands.* 1947–1973.

Washington State Department of Natural Resources. 1990.

White, C. A. *A History of the Rectangular Survey System.* U.S. Department of Interior Bureau of Land Management. 1980.

## Chapter 13

Black, H. C. *Black's Law Dictionary.* St. Paul, MN: West Publishing Co., 1979.

Brown, C. M. *Boundary Control and Legal Principles.* New York: John Wiley & Sons, 1969.

Brown, C. M., and Eldridge, W. H. *Evidence and Procedures for Boundary Location.* New York: John Wiley & Sons, 1967.

Grimes, John S. *Clark on Surveying and Boundaries.* Charlottesville, VA: The Michie Co., 1976.

**Chapter 14**

*Definitions of Surveying and Associated Terms.* American Society of Civil Engineers and the American Congress on Surveying and Mapping. 1972.

Pafford, William A. *Handbook of Survey Notekeeping.* New York: John Wiley & Sons, 1962.

U.S. Naval Education and Training Command. *Engineering Aid 3 & 2, Vol. 1.* 1976.

Wattles, Gurdon H. *Survey Drafting.* Orange, CA: Gurdon H. Wattles Publications, 1977.

**Glossary**

*Definitions of Surveying And Associated Terms.* American Society of Civil Engineers and the American Congress on Surveying and Mapping. 1972.

# Index